Northwest Mountain Weather
Understanding and Forecasting
for the Backcountry User

Northwest Mountain Weather
Understanding and Forecasting for the Backcountry User

Jeff Renner

THE MOUNTAINEERS

Published by The Mountaineers
1011 SW Klickitat Way, Seattle, Washington 98134

Published simultaneously in Canada by Douglas & McIntyre, Ltd., 1615 Venables Street, Vancouver, B.C. V5L 2H1

Published simultaneously in Great Britain by Cordee, 3a DeMontfort Street, Leicester, England, LE1 7HD

Manufactured in the United States of America

Printed on recycled paper

Edited by Barry Foy
Illustrations by Nick Gregoric
Cover photograph: Mount Rainier, by Pat O'Hara
Cover design by Elizabeth Watson
Book design by Bridget Culligan
Composition by Scribe Typography

Library of Congress Cataloging in Publication Data

Renner, Jeff.
    Northwest mountain weather : understanding and forecasting for the backcountry user / Jeff Renner.
        p.   cm.
    Includes bibliographical references and index.
    ISBN 0-89886-297-3
    1. Northwest, Pacific—Climate—Handbooks, manuals, etc.
2. Mountains—Northwest, Pacific—Recreational use—Handbooks, manuals, etc. 3. Weather forecasting—Northwest, Pacific—Handbooks, manuals, etc.
I. Title.
QC984.N97R46   1992
551.65795—dc20   91-46559
                                                                        CIP

Contents

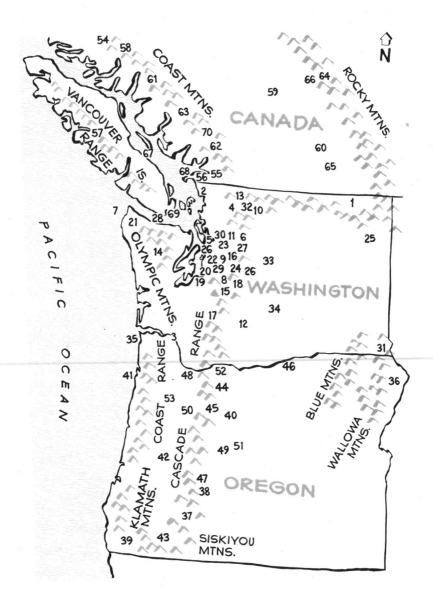

Figure 1. *The Pacific Northwest.*

Washington

1. Abercrombie Mountain
2. Bellingham
3. Columbia River
4. Diablo Dam
5. Everett
6. Glacier Peak
7. Hoh Head
8. Huckleberry Mountain
9. Issaquah
10. Liberty Bell
11. Monte Cristo
12. Mount Adams
13. Mount Baker
14. Mount Olympus
15. Mount Rainier
16. Mount Si
17. Mount St. Helens
18. Naches Pass
19. Olympia
20. Puget Sound
21. Quillayute
22. Seattle
23. Skykomish
24. Snoqualmie Pass
25. Spokane
26. Stampede Pass
27. Stevens Pass
28. Strait of Juan de Fuca
29. Tacoma
30. Verlot
31. Walla Walla
32. Washington Pass
33. Wenatchee
34. Yakima

Oregon

35. Astoria
36. Baker
37. Crater Lake
38. Diamond Peak
39. Klamath Mountains
40. Lake Chinook
41. Lincoln City
42. Mary's Peak
43. Medford
44. Mount Hood
45. Mount Jefferson
46. Pendleton
47. Pengra Pass
48. Portland
49. Redmond
50. Salem
51. Smith Rock
52. The Dalles
53. Willamette Valley

British Columbia

54. Franklin Glacier
55. Fraser Plateau
56. Fraser River
57. Golden Hinde
58. Homathko Icefield
59. Kamloops
60. Kelowna
61. March Mountain
62. Mount Garibaldi
63. Mount Gilbert
64. Mount Revelstoke
65. Penticton
66. Revelstoke
67. Strait of Georgia
68. Vancouver
69. Victoria
70. Whistler

Preface

◆

*D*uring the late winter and spring or 1980, when Mount St. Helens was grabbing headlines, I spent most of my time in the Cascades near the volcano. As Science Editor for Seattle's KING Television, I was assigned to cover the rumblings of the once-again active volcano.

Our crew spent daytime hours bouncing up and down logging roads in our newscar (the shock absorbers were a loss after the first week) or flying from summit to glacier to Spirit Lake in the station's helicopter. The evening hours were devoted to transmitting live broadcasts from a variety of lookouts. At night we camped out—first in a van jammed with electronic and photographic gear, reference texts, and food and clothing on a ridge about 12 miles northwest of St. Helens, then in tents clustered on a mountaintop just 5 miles from the volcano.

We quickly learned about mountain weather firsthand. The stars of one hour could give way to a pelting of hailstones the next (and, eventually, pumice from the erupting volcano). Obtaining our much-needed sleep when 50- and 60-mile-per-hour winds buffeted our fragile quarters wasn't easy.

When good sense dictated we leave our campsite for the last time (less than 48 hours before the big eruption), I had gained a deep appreciation for the force and variety of mountain weather in the Pacific Northwest, to say nothing of the power of volcanoes. That appreciation would be expanded by later climbing and hiking trips throughout the Cascades, Olympics, and Coast mountains.

When my duties were expanded to full-time television meteorologist, I found myself hearing reports of weather-related climbing and hiking accidents, all too often due to a lack of basic weather knowledge, inadequate pre-trip planning, or insufficient vigilance on the mountain.

With my background as a commercial pilot and flight instructor, I began to think that mountain travelers could avoid many accidents by using some of the same techniques to assess and monitor the weather before and during a flight. That's what gave birth to this book.

I've worked to distill the essentials of mountain weather in the first five chapters, emphasizing how they apply to climbing, hiking, skiing, and snowshoeing. Chapter 6 points you toward sources of weather information, and guides you through the gathering of the data and forecasts important to your planned trip. You'll find concrete outlines and examples of how to analyze that information, to make good "go" or "no go" decisions. Chapter 7 is a field guide within a guide, a series of checklists to cover a wide variety of situations. Organized in a decision-tree format, it will quickly guide you to check the key information you will need.

It would be easy to view this book solely as a guide to avoiding risky weather and hazardous situations in the mountains. I'd like to raise the focus a little: it's also written to help you find the greatest enjoyment in our beautiful environment. I look forward to seeing you on the trail.

Acknowledgments

Although there is one name on the cover of this book, in reality it represents the efforts and influence of many. Thanks are due to Mark Anderson, my longtime friend and colleague, who introduced me to the mountains of the Northwest and worked with me as a photographer on Mount St. Helens. Rich Marriott, a friend, colleague, and first-rate meteorologist and avalanche forecaster, provided invaluable guidance, particularly for the sections on snow and avalanche conditions. Pam Speers-Hayes graciously agreed to allow the use of part of her master's thesis on precipitation patterns in the Cascades and Olympics.

Appreciation is also due the faculty of the Department of Atmospheric Sciences at the University of Washington, where I received my degree. The research of many, in particular Cliff Mass, Richard Reed, Peter Hobbs, and Mark Albright, have clarified many of the previously mysterious weather patterns of this region.

I would also like to thank Mountaineers Books and in particular my editor Don Graydon for the patience and persistence needed to guide a first-time author from concept to completed book. Finally, special thanks to my wife Sue and son Eric for their forbearance of the hours spent working on this project, and for their company in doing "field research." Also, to my parents for their enthusiastic support.

Jeff Renner
Redmond, Washington

A note about safety

Safety is an important concern in all outdoor activities. No book can alert you to every hazard or anticipate the limitations of every reader. Therefore, the descriptions in this book are not representations that a particular day, place, or excursion will be safe for your party. When you engage in outdoor activities, you assume responsibility for your own safety. Under normal conditions, such excursions require the usual attention to traffic, road and trail conditions, weather, terrain, the capabilities of your party, and other factors. Keeping informed on current conditions and exercising common sense are the keys to a safe, enjoyable outing.

The Mountaineers

CHAPTER 1

Climate and Weather of Pacific Northwest Mountains

◆

Two groups of climbers are setting up camp on Mount Baker, in Washington's North Cascades. A cold front moved through yesterday, and the rain-washed sky is a brilliant blue, dotted with small, white cumulus clouds. Neither group notices the white puffs growing into massive, dark-gray thunderheads until the clouds begin to merge, cutting off the sunlight. Both groups are surprised. Neither listened to updated forecasts warning of a second disturbance that would lead to showers, especially along the western slopes of the Cascades.

It begins to pour, a chill rain that is soon mixed with driving ice-pellet showers, snowflakes, and thunder. Both groups reluctantly pack up their gear. The first drives west, back down the mountain to Bellingham, giving up hope of climbing. Relative newcomers to the area, they aren't acquainted with the wide variation in weather patterns in the mountains of the Pacific Northwest. The second group, realizing they should have kept a closer eye on the forecast, heads east over the North Cascade Highway. Once they get over Washington Pass, the skies clear. The planned climb of Mount Baker is replaced with climbs of Liberty Bell and surrounding peaks.

The Pacific Northwest is a region born of the fire of volcanism and sculpted by water and ice. The activity of Mount St. Helens is just one proof that this process of building and eroding continues, that the mountains of the Pacific Northwest remain geological adolescents. Understanding the geology of this region is essential to understanding its weather patterns, and understanding its weather patterns is essential to planning trips into the mountains for hiking, climbing, and skiing.

The mountains of the Pacific Northwest, like those elsewhere, were created by geological forces acting deep beneath the earth's surface. Our planet's continents and ocean floors are really a collection of picture puzzle–like pieces known as *plates.* These plates float on rock that is partially melted in the upper region of the earth's interior, called the mantle, and fully melted deeper within this layer. Temperatures in the mantle may range from 2,000 to 3,400 degrees Fahrenheit (1,100°–1,900°C). This melted rock flows in currents as slow as cold molasses, which drive the massive plates, inch by inch.

In the Pacific Northwest, a plate making up part of the Pacific Ocean floor is colliding with and diving beneath the lighter plate that makes up our continent. Because the boundary between these colliding plates runs

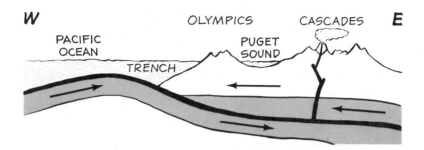

Fig. 2. *Mountain- and volcano-building in the Northwest.*

essentially north-south, the major mountain ranges thrust up by the collision also run essentially north-south (fig. 2).

Visitors flying into the Pacific Northwest on a clear day enjoy a spectacular sight: the blue waters of the Pacific giving way to a rugged coastline marked by the Coast Range of Washington and Oregon, interrupted by the broad Columbia River and then the Chehalis River Gap in southwestern Washington. The young Olympic Mountains rise to the north, separated from British Columbia's Vancouver Island Range by the Strait of Juan de Fuca, which is the shipping thoroughfare from the Pacific into Puget Sound and Howe Sound. The Vancouver Island Range is the backbone of Vancouver Island.

A large trough runs along the east side of these ranges, beginning with Oregon's Willamette Valley and continuing into the Puget Sound Basin in Washington and then Hecate Strait and the Strait of Georgia in British Columbia.Farther east, the land rises again to form the Cascade Range, which extends from northern California through Oregon and Washington, then merges with British Columbia's Coast Mountains. There are two great valleys that carve their way through the Cascades: that of the Columbia River in Washington and Oregon, and the Fraser River Valley in British Columbia.

These mountain ranges split Washington, Oregon, and British Columbia into distinctly different climatic zones — a fact used to good advantage by the knowledgeable climbers in the story at the beginning of this chapter. The differences in climate and the resulting variations in plant and animal life are consequences of the way the mountain ranges interact with the region's major weather patterns.

Variations in Precipitation

Because most of the storms that move into the Pacific Northwest form over the Pacific Ocean and are directed eastward by winds high in the atmosphere, they end up on a collision course with these mountain ranges. As the moisture-laden air collides with the mountains, it's forced to rise. The resulting cooling triggers a massive growth spurt in the clouds and a large increase in precipitation along the mountains' western slopes. The process is similar to wringing out a soggy sponge; the effect is certainly the same.

Along the moisture-rich western slopes, hiking trails wind through towering stands of western hemlock, Douglas fir, and western red cedar. The often spongy forest floor is an ideal habitat for the aptly named banana slug, much to the delight of youngsters and the disgust of adults.

Just over the passes, on the eastern side of the mountains, a startling transformation occurs. Widely spaced, drought-tolerant lodgepole pines replace the firs. The forest floor is often rock-hard, with sparse vegetation, and the ledge that seems to offer a superb handhold may serve as an equally superb sun deck for a rattlesnake.

The reason for the stark contrast is simple. Once over the crest, the air, already relieved of some of its moisture, dries further as it descends the eastern slopes and warms. The result is a dramatic variation in precipitation, even over short distances (fig. 3). Stampede Pass, in the central Washington Cascades, receives 91 inches (231 cm) a year. Yakima, to the east and in the shadow of the Cascades, averages only 8 inches (20 cm) — that's less than 10 percent of what falls on Stampede Pass, which is only 64 miles (102 km) away. Variations of this magnitude are unheard of in other regions of the United States and Canada, but are common in the Pacific Northwest.

Forcing moist air up hills just 200 feet (60 m) higher than the surrounding terrain can double or even triple short-term precipitation (over 12 hours or less), and the Cascade crest ranges in elevation from 4,000 to 9,000 feet (1,219–2,743 m). But large variations in precipitation amounts also occur among sites at the same elevation, often due to differences in slope and exposure.

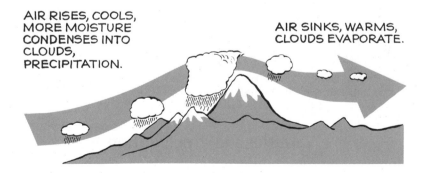

Fig. 3. *Influence of mountains on precipitation.*

Precipitation in British Columbia

The Vancouver Island Range, which forms the backbone of British Columbia's Vancouver Island, produces strong contrasts in weather. Precipitation rates increase rapidly upslope, exceeding 125 inches (317 cm) on Mount Modeste and Mount Todd. The rate drops to less than 30 inches (76 cm) on some of the Gulf Islands — it's no wonder this area is called the Sunshine Coast.

Farther east, Vancouver receives an average of 47 inches (119 cm) of precipitation each year. Precipitation amounts increase rapidly with elevation to the east, with yearly averages of 95 inches (241 cm) near the crest of the Coast Mountains. Descending the eastern slopes of the range and out onto the Fraser Plateau, the average drops off rapidly, to as little as 16 inches (41 cm) a year.

Precipitation in Washington

In Washington, the variations are equally impressive. The Coast Range boosts average annual precipitation to as much as 120 inches (304 cm) just east of Willapa Bay in southwestern Washington, with a rapid drop to only 40 inches (101 cm) in Chehalis.

By far the most dramatic variation, though, is seen in and around the Olympic Mountains. Hoh Head, along the rugged northern Washington coast, averages 90 inches (228 cm) a year. Mount Olympus, only 35 miles (56 km) to the east, averages 240 inches (609 cm) near its summit.

Moving east to the Puget Sound basin, Seattle receives an average of roughly 36 inches (91 cm) a year. Issaquah, a suburb in the foothills of the Cascades, receives 50 (127 cm). Yearly averages of more than 100 inches (254 cm) are not uncommon along the Cascades to the east, with as much as 190 inches (482 cm) near Monte Cristo.

The dry east side of the Washington Cascades is truly dry, often receiving 10 inches (25 cm) of rain or less each year.

Precipitation gauges in the Cascades show large variations even for stations at similar elevations and on the same side of the range. Hourly amounts from Verlot south to Stampede Pass are three to four times those measured from Stampede Pass south to Mount Rainier. The path of storms, slope exposure, and the protection offered by the smaller coastal mountains are responsible for these additional differences.

Precipitation in Oregon

Oregon also produces startling variations in precipitation. Annual rates increase abruptly just east of Lincoln City along the coast, from 70 inches (177 cm) along the beaches to as much as 200 inches (508 cm)

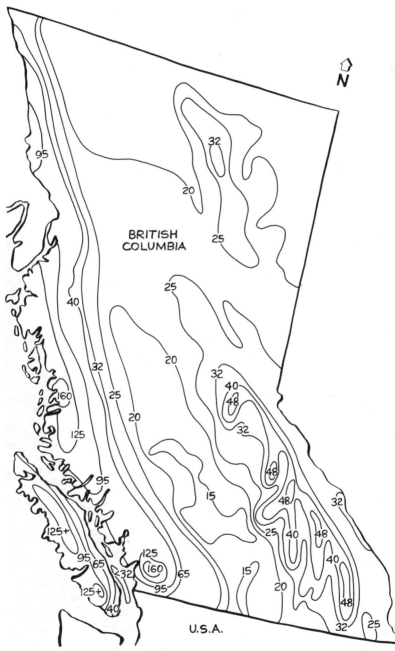

Fig 4. *Average annual precipitation in inches, British Columbia (adapted from the* Climatic Atlas of Canada*).*

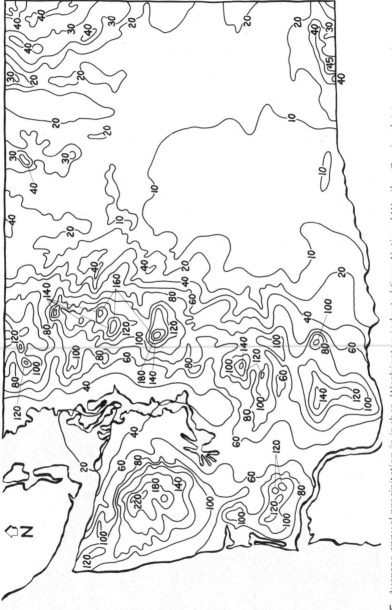

Fig. 5. *Average annual precipitation in inches, Washington (adapted from National Weather Service data).*

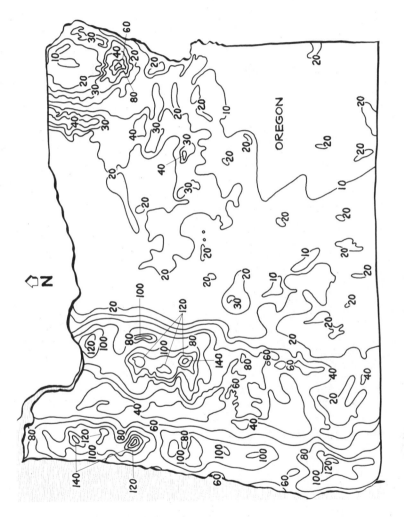

Fig. 6. *Average annual precipitation in inches, Oregon (adapted from National Weather Service data).*

near the summits of the Coast Range, then dropping abruptly to just 40 inches (101 cm) in Salem, sheltered in the Willamette Valley.

Suburban Hillsboro, to the west of Portland, receives an average of 38 inches (97 cm) per year. Troutdale, to the east of Portland, averages 48 (121 cm). Ascending Mount Hood, that spectacular volcano and precipitation-catcher to the east, precipitation averages soar to 130 inches (330 cm) per year. Once over the Cascades, amounts decline rapidly. The Dalles, along the Columbia River, averages less than 15 inches (38 cm).

Precipitation in the Rockies and neighboring ranges

Much smaller but still significant variations exist to the east of the Cascade/Coast Mountains region. The Blue Mountains stretch from the central Oregon Cascades through northeastern Oregon and southeastern Washington to the mountains of central Idaho. Precipitation averages increase from 15 inches (38 cm) per year in Walla Walla to more than 40 inches (101 cm) at Oregon Butte and Diamond Peak just to the east-northeast.

Concluding this survey of Pacific Northwest mountain geography and precipitation is the northern extension of the Rocky Mountains, which ranges from Idaho through northeastern Washington and into British Columbia and Alberta. Wet zones tend to correspond to the highest peaks along the western edge of the Rockies. Sherman Peak in central Idaho averages more than 32 inches (81 cm) per year. Abercrombie Mountain in northeastern Washington wrings out an average of 45 inches (114 cm) from passing clouds, as does Mount Revelstoke in southeastern British Columbia.

In general, maximum precipitation along the western slopes of the Rockies tends to be three to four times that found in the Fraser Plateau in British Columbia and the Columbia Basin and High Lava Plains of eastern Washington and Oregon. This is a significant increase in precipitation, but nowhere near that found along the ranges that border the coastlines of Washington, Oregon, and British Columbia. Most of the major weather systems move in off the Pacific and have dropped much of their soggy load by the time they reach the Rockies.

— Different Storm Tracks, Different Weather —

Because the Pacific is the birthplace of most of the storms that hit this region, forecasters pay close attention to the *source region* of the weather systems that develop. Weather systems moving from one area of the Pacific may produce prolonged, heavy rains that soak both eastern and western slopes; those from another area might generate intermittent showers, confined to the western slopes. Know where a storm was formed, and the path it has traveled (its *storm track*), and you have a quick way to assess its likely impact and to choose the destination offering the best weather for mountain recreation.

From South-Southwest

The Pineapple Express, true to its name, tends to direct warm and very humid air up from the subtropics or even the tropics (fig. 7). We can often see a band of clouds stretching from the Hawaiian Islands to Vancouver Island in satellite pictures. Because the temperature and moisture content of this air are very high, we get rain when it collides with the

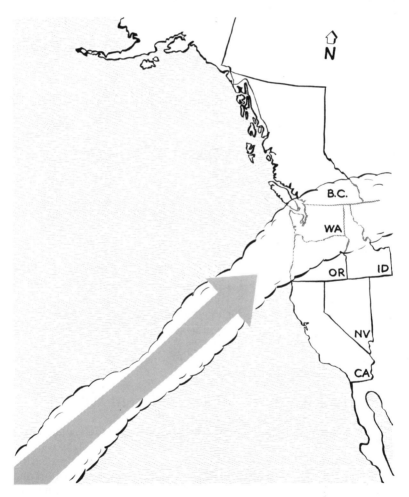

Fig. 7. *South–southwest storm track.*

much cooler air in the Pacific Northwest and is lifted.

The Pineapple Express often brings 1 to 2 inches (2.5–5.1 cm) to places like Puget Sound and the Willamette Valley, and 3 to 4 inches (7.6–10.2 cm) to the western slopes of the Cascades, Olympics, and the Coast Mountains of British Columbia. This storm track tends to persist for several days as new disturbances develop and ripple along the band of clouds.

The very mild air moving up from the tropics usually raises the freezing level to 8,000 feet (2,438 m) or higher. Snow melts rapidly in the mountains, often leading to serious flooding. The air is usually very stable, so thundershowers are less common.

From Southwest or West

A more common storm track is from the southwest to west (fig. 8). The temperature and moisture content of the air are usually lower; the result is less-intense (although not necessarily light) rainfall. The rain isn't as persistent as with the Pineapple Express.

Although the prevailing westerlies can fire off one disturbance after another at us, there is usually a break between them, sometimes lasting only a few hours, occasionally a full day. Cooler air moves in after each

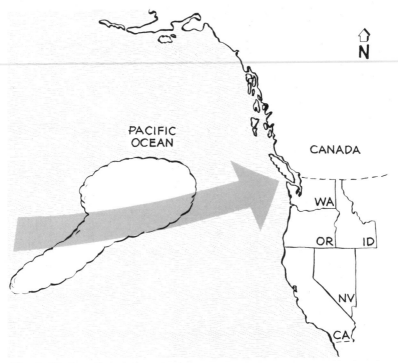

Fig. 8. *West–southwest storm track.*

22

disturbance, hence a better chance of thundershowers, especially on the windward slopes of mountains, where the air is given an additional thrust upward.

If cold air has an icy grip on the Northwest prior to the arrival of a system in this storm track, the precipitation may start out as snow but gradually change to rain. Freezing levels associated with this pattern vary from season to season, from 3,000–5,000 feet (914–1,524 m) in winter to 5,000–7,000 feet (1,524–2,133 m) during autumn and spring. The result is slushy, wet snow at or near pass elevations.

From the West to Northwest

Precipitation from this storm track (fig. 9) usually doesn't last long, but it can come as a rude surprise. When the storm track is from the northwest, with more prolonged movement over the Pacific, weather systems

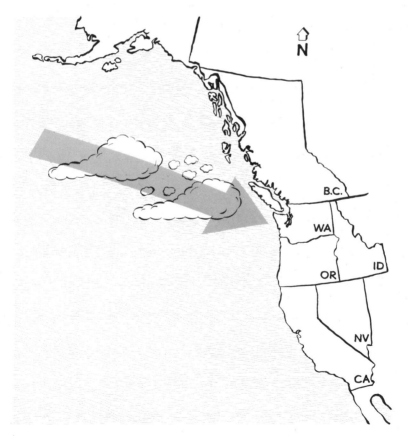

Fig. 9. *West–northwest storm track.*

move through quickly, with fairly rapid clearing. However, the contrast between temperatures in this pattern can lead to thundershowers, especially along the western slopes of the Cascades. Expect freezing levels from just above sea level to 3,000 feet (914 m) in winter, to 3,000–5,000 feet (914–1,524 m) in autumn and spring. This pattern produces excellent-quality snow in the mountains, and usually lots of it.

From the North

A storm track from the north (figs. 10 and 11) is this region's most frequent producer of snow in winter in coastal British Columbia, the Puget Sound lowlands, and the Willamette Valley.

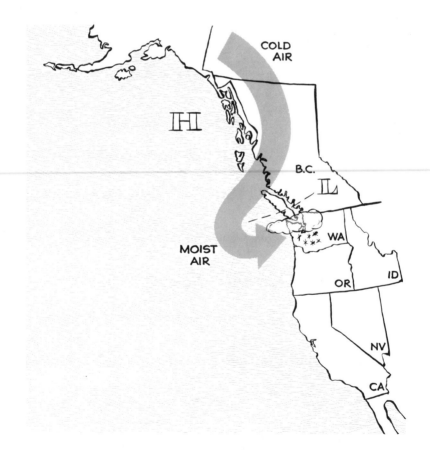

Fig. 10. *Northerly storm track.*

24

Cold air is drawn from the interior of British Columbia, or the Yukon, and circulated over the Pacific just long enough to pick up moisture, but not long enough to warm the air so that rain is produced instead of snow. As the disturbances drop southward along the coast, snow makes its way from Vancouver to Bellingham to Seattle and sometimes Portland. If snow is falling in the cities, it is almost certainly falling in the mountains; but because the northerly winds slide along, and not against, the major

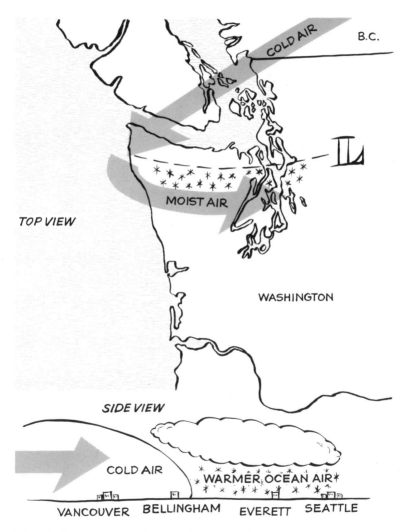

Fig. 11. *Snow pattern with northerly storm track.*

25

Northwest mountain ranges, more snow may fall in the flatlands than in the mountains. This pattern, however, is not terribly long-lived, usually not more than one day.

—— Latitude and Weather in the Northwest ——

Because the *jet stream,* that fast-moving river of air high in the atmosphere that directs storm tracks, is strongest and farthest south in the winter months, that's when the Pacific Northwest receives most of its precipitation. Both Seattle and Portland receive less than 10 percent of their annual precipitation during the summer months of June, July, and August. Knowing this pattern can assist you in picking the best time of year for extended hiking, climbing, or skiing trips.

The northerly retreat of the jet stream during the summer doesn't benefit all areas equally. Although it tends to remain over southern Alaska and northern British Columbia, weak disturbances still move far enough south to make the North Cascades of Washington, the Olympics, and the Coast Mountains of British Columbia markedly cloudier and wetter than the Cascades from central Washington southward.

———— Mountains and Temperature ————

The Cascades of Washington and Oregon, and British Columbia's Coast Mountains, create more than wet and dry zones; they also play a major role in producing variations in temperature. The mountains tend to confine the moist ocean air to the west, which has a moderating effect.

During the summer months, the mountains' role as a barrier to moist ocean air permits temperatures from the Cascade crest eastward to far exceed those to the west. The average July daytime high temperature in Seattle is 75 degrees Fahrenheit (24°C), while in Yakima it's 88 (31°C).

From October through March, the Cascades and British Columbia Coast Mountains usually deflect to the east the bitterly cold arctic outbreaks from Alaska and the Yukon and Northwest territories. The average January overnight low temperature in Seattle, for example, is a relatively mild 34 degrees (1°C), while in Yakima it's 18 (–8°C).

This diversity in our mountains is what makes our region so special and enjoying them on foot, on skis, or on snowshoes such a delight. It's all a consequence of the geological forces that shaped the Pacific Northwest, and the region's proximity to the weather factory of the Pacific Ocean. The result is not one climate zone but several, each tending to run north-south. Visualizing the big picture is important to understanding the more subtle details of Pacific Northwest weather, and using that understanding to make trips into the mountains enjoyable and safe.

CHAPTER 2

What Makes It Wet

◆

We should have left our skis in the garage. Recently fallen snow is dripping from tree limbs, big drops splashing into the puddles of slush lining Interstate 90 on the way to Snoqualmie Pass in Washington. Soft curtains of fog veil approaching cars, their drivers discouraged and returning home. They, like us, ignored the long fingers of cirrus clouds yesterday morning that broadened into blanketlike sheets of gray stratus by sunset. Pressing on with the unshakable optimism unique to early-season cross-country skiers, we are arguing over which, if any, wax will salvage the day when we notice the fog thickening, obscuring the line of cars stopped just ahead. Sliding to a swerving halt in snow the consistency of runny oatmeal, we reluctantly agree to call it quits, and head for home.

A little attention to the thickening, lowering clouds moving in from the southwest in the episode described above could have saved us a long drive and disappointing snow conditions. That change in cloud patterns signaled the approach of warmer air and precipitation.

At times, weather systems in the Northwest cover the entire region, and the choice is a simple one: stay home. More frequently, patterns in and near the mountains are subtle and localized. While deciphering the clues can be difficult, such patterns offer hikers, climbers, snowshoers, and skiers more options than a simple go or no-go. As a step toward understanding these patterns, let's begin with some basics.

A Little Basic Meteorology

Temperature

The sun is the engine that drives our atmosphere. It provides the heating that, together with several other factors, creates the temperature variations that are ultimately responsible for wind, rain, and snow.

The earth's location — 93 million miles (148,800,000 km) from the sun — is what makes life as we know it possible. Venus, with an orbit closer to the sun, experiences average surface temperatures of roughly 800 degrees Fahrenheit (426°C), while the more distant Mars averages 81 degrees below zero (–63°C).

Proximity to the sun is only one factor. The intensity of the sun's radiation varies across the earth's surface. Given a choice between Mount McKinley in Alaska and Crater Lake in Oregon for an autumn backpacking trip, for instance, a hiker with limited tolerance for cold temperatures will likely choose Crater Lake. That's because Crater Lake is closer to the equator; the sun will be more directly overhead at noon, and therefore the heating from the sun will be more intense.

This relationship between heating from the sun and the angle of the sun above the horizon also explains why summer is warmer than winter: the sun is more directly overhead. You can see how this works by shining a flashlight on this page, first from directly overhead, then at an angle. The beam of light shining from directly above the page has a smaller area to illuminate and heat than a beam striking the surface at an angle. The smaller the area illuminated by the flashlight (or the sun), the more intense the heating.

Given more intense sunlight at the equator than at the poles, the temperature differences come as little surprise. But extremes in temperature, large as they may be, are controlled by the movement of air. Differences in air temperature lead to air movement, which prevents runaway heating or cooling.

Air Pressure

Anyone who's had to chase a tent in a windstorm knows that air moves sideways. But it also rises and descends, movement that can generate or dissipate clouds.

When air rises due to heating, it's as if it were shedding extra pounds. Air has weight, and just as the reading on our bathroom scale drops when we lose weight, the reading on a barometer, which measures air pressure, falls when some of the air moves up and away.

Just as heating air makes it rise, cooling it will make it sink. Because cold air is more dense than warm air, it tends to find its way to the bottom of the atmosphere, that is, to the ground. Cold air tends to collect in low places such as valleys and canyons, making them chilly campsites on cold, windless nights.

To summarize, the sinking of cool air increases air pressure, while the rising of warm air decreases it. These pressure differences, the result of temperature differences, produce moving air, which we refer to as wind. Air will generally move from an area of high pressure to one of low pressure.

Clouds

Air moving from high to low pressure carries moisture with it. As that air cools, as a result of either rising or moving over a colder surface, the moisture condenses into clouds or fog. This occurs because as air cools, its capacity to hold water vapor is reduced. For example, air at 98.6 degrees Fahrenheit (37°C), our body temperature, is capable of holding roughly thirteen times as much water vapor as it can at 30 degrees (–1.1°C).

When that moisture-laden air is cooled, then, its capacity to hold water vapor is rapidly reduced. Not all of the water vapor will "fit," and that which "spills out" condenses into a cloud of water droplets.

When air cannot hold any additional water vapor, we say it is saturated. Meteorologists call that saturation point the *dew point.* The dew point is simply the temperature at which the air will become saturated with moisture as the air cools, and clouds will usually form. We encounter a similar effect when we see our breath on a cold day. As we inhale, our body warms the air to approximately 98.6 degrees Fahrenheit and adds moisture. As we exhale, that warm, moist air is cooled by the colder air around us, leading to condensation of the water vapor (a gas) into water droplets (a liquid).

Therefore, the dew point is always equal to or cooler than the air temperature, never warmer. When the air temperature cools to the dew point, water vapor condenses into water droplets, and clouds or fog form. No clouds will form when the temperature is much higher than the dew point.

Relative humidity compares how much water vapor the air is holding with how much it could hold. Think of it as a measure of how saturated the air is. Relative humidity is usually expressed as a percentage; 75 percent relative humidity, for example, means the air is holding three-quarters of the water vapor it's capable of holding.

The process of cooling and condensation operates on a large scale in the atmosphere as air moves from high-pressure into low-pressure systems and is lifted. The result is, at times, a weather system that covers the entire Northwest, though mountain weather patterns are more often localized. Let's look now at large weather systems.

Large-Scale Weather Systems

Experienced mountain travelers know that precipitation along the western slopes of the mountain ranges in the Pacific Northwest isn't always light and often encompasses the entire region in a soggy embrace. The cause? Large-scale weather systems, usually moving in from the Pacific.

The Gulf of Alaska is the factory for most of the storms that batter the Pacific Northwest. This is a natural consequence of its latitude and geography. The gulf is the battleground between air moving south from the Arctic and air moving north from the subtropics and midlatitudes, which gives birth to many of the storms that affect our recreational plans in the mountains.

Because polar and arctic air is colder and therefore more dense than air farther south, it sinks. The zone where it sinks and "piles up" is a region of *high pressure.* As the air sinks and its pressure increases, its temperature also increases. The effect is similar to what happens to football players caught at the bottom of a pile. The players on the bottom get squeezed the most, and their temperature (and possibly their temperament!) heats up. In the atmosphere, this warming within a high tends to evaporate the little moisture present in cold polar and arctic air. This is why the Arctic is classified as a desert, receiving very little precipitation — not all deserts are covered by sand!

Barrow, for example, on the north slope of Alaska, receives an average of only 28 inches (71 cm) of snow each year, in contrast to Fairbanks, which receives 66 (167 cm), and Juneau, which averages 105 inches (266 cm). Each town is progressively farther south.

If our planet didn't rotate, this cold air would just continue to slide southward to the equator. Intense solar heating near the equator forces air to rise, creating a region of *low pressure* that rings the globe. Because air within this band rises, it also cools, which tends to condense water vapor into droplets that form clouds, just as your breath condenses on a

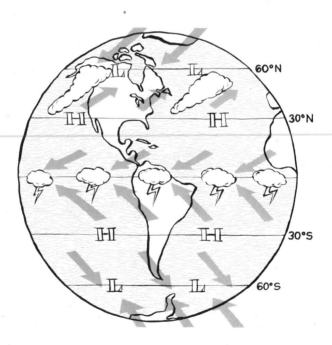

Fig. 12. *Circulation patterns.*

cold day. Satellite pictures show a series of thunderstorms marking this equatorial low-pressure zone, which is called the Intertropical Convergence Zone. It's a very wet area: more rain can fall in a single day within this zone than during an entire month in Washington, Oregon, or British Columbia.

But the air sinking and moving south from the pole and that rising from the equator don't form a simple loop moving from north to south and back again. The rotation of the earth is responsible for deflecting this air, creating a considerably more complicated circulation of air over our planet (fig. 12).

Some of the air rising from the equator descends over the subtropics. This sinking air creates a region of high pressure. As it sinks within this high, the earth's rotation produces a force known as the *Coriolis force*. The combination of the Coriolis force and the effect of friction from air moving over land and water causes the air sinking within the high-pressure system to rotate in a clockwise direction in the northern hemisphere, and counterclockwise in the southern hemisphere.

Fronts

Some of the air that sinks and spreads outward from these subtropical highs picks up moisture from the oceans and moves north, eventually meeting the cold, dry air spreading southward from the pole. The boundary between these two very different types of air masses is called the *polar front,* and it rings the globe in both hemispheres.

When this boundary between different air masses doesn't move, it's also called a *stationary front.* In the Gulf of Alaska and elsewhere, it serves as a nursery for the development of storms.

During the autumn and winter months, the air moving south from the Arctic toward the Gulf of Alaska can be as cold as 40 or 50 degrees below zero (–40°/–45°C). The temperature of the air over the gulf is moderated by water's capacity to absorb and retain heat. Air temperatures there may be 30 to 40 degrees above zero (–1.1° to –4.4°C), yielding an impressive contrast along this polar front of as much as 90 degrees Fahrenheit.

Because of this great contrast in temperatures, the polar front is especially strong in and around the Gulf of Alaska. But it rarely remains stationary there, or anyplace else. Imbalances caused by the rotation of the earth and the differing influences of land, sea, ice, and mountains allow the cold, dry, dense air from the north to slide south, forcing some of the warm air to rise. The zone where the cold air is replacing the warm air is referred to as a *cold front* (fig. 13).

Conversely, farther east, warm air is forced to glide up and over the

31

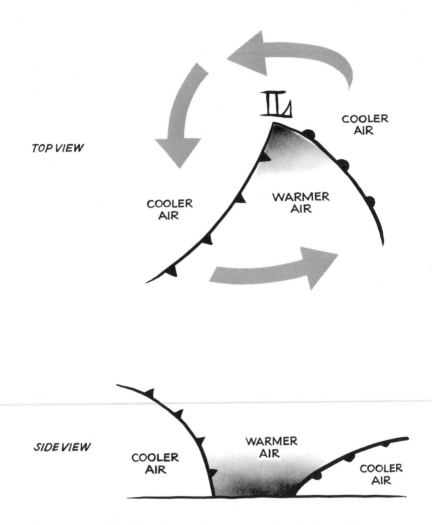

Fig. 13. *Low with cold and warm fronts: top, top view; bottom, side view.*

cooler air near the surface. This zone where warm air is gradually re-placing cooler air is referred to as a *warm front* (fig. 13). This "wave" or bend on the stationary front may develop into a low-pressure system, with air circulating counterclockwise around the low, exactly the oppo-site direction of air moving around a high — again a consequence of the earth's rotation and friction.

When low-pressure systems develop in the Gulf of Alaska, the coun-terclockwise circulation of air around the low draws warm, moist air northward from over the Pacific Ocean, and very cold air southward from the interior of Alaska.

Clouds

Different types of clouds develop along cold and warm fronts, and it's the differences in their shape that can provide the field forecaster with valuable clues to coming changes in the weather. Variations in cloud shapes along warm and cold fronts mirror differences in the physical processes that are taking place along those fronts.

Warm-Front Clouds

If a low-pressure system is moving across the Pacific, the first clouds visible in the Pacific Northwest are those associated with the warm front. Because the warmer air moving inland is less dense than the cooler air it's replacing, it is lifted, gradually sliding up and over the cooler air at the earth's surface. As that warm, moist air from the Pacific rises, it cools. If this rising air cools to its dew point, the water vapor in the air mass will either condense into water droplets or *sublime* into ice crystals.

Sublimation is the atmosphere's shortcut, transforming water vapor directly into ice crystals without it first condensing into liquid water. The result is determined by the temperatures aloft. Because the advancing warm air usually rises very high, 20,000 feet (6,096 m) or more, the temperatures are well below freezing, certainly cold enough to sublime the water vapor into ice crystals.

Cirrus Clouds

The first clouds we see as the warm front advances are the fibrous high ones called *cirrus clouds.* They are thin, often less than 1,000 feet (304 m) thick.

The clouds' ice crystals or water droplets act as miniature prisms, bending and splitting sunlight or moonlight into its component colors. The result is the halo often seen ringing the sun or moon as cirrus clouds move in with an approaching warm front. Such halos are very wide and change in color from red at their inner ring to yellow to green to blue; they usually precede precipitation by 24 to 48 hours.

Lenticular Clouds

Another type of cloud is created by moist air moving up and over a mountain; we see it frequently in the Northwest, especially above volcanic peaks such as Mount Rainier or Mount Hood. It's called a *lenticular cloud,* that is, a cloud shaped like a lens (fig. 14). Such clouds are often hints that a weather disturbance is nearby, and that a warm front may be approaching.

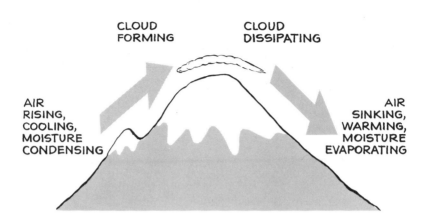

Fig. 14. *Mountain lenticular cloud.*

Lenticular clouds are formed when moisture high in the atmosphere is deflected upward when it runs into a major peak. As that moist air rises, it cools sufficiently to condense into a cloud. But as it passes over the peak and begins descending, it warms, and the water droplets or ice crystals that make up the lenticular cloud evaporate. Although lenticulars appear to be stationary (they've often been mistaken for UFOs), they are continually dissipating on the leeward edge. If followed by cirrus and eventually stratus clouds, such lenticulars often give climbers, backpackers, and skiers 24 to 48 hours' notice of approaching precipitation.

Lee-Wave Clouds

The up-and-down motion of air traveling over a mountain peak or range occasionally continues to the downwind side, much like the ripples produced by a rock thrown into a pond. This results in a series of clouds forming and dissipating as the air rises and falls in wavelike fashion.

Such clouds are called *lee-wave clouds* and can stretch for hundreds of miles downwind of the mountain barrier. Because these clouds form to the lee side of mountain ranges, they are most frequently found to the east of the Cascades in the Northwest. When oriented properly to the sun, wave clouds are often tinted in beautiful, iridescent colors.

Stratus Clouds

As the warm front advances, the boundary between the warm air and the cooler air below gradually lowers closer to the ground. More moisture is available at the warmer temperatures found closer to sea level, and thicker clouds form: sheetlike *altostratus clouds,* ranging from 20,000 to as little as 6,000 feet (6,096–828 m) above sea level, and thick, blanketlike layers of *stratus clouds,* whose thickness can range from a few hundred feet to several thousand.

Another circular "rainbow" of sorts often rings the sun or moon through altostratus. This is called a *corona,* and it hugs the sun or moon much more closely than a halo. For that reason, while halos tend to indicate that precipitation is at least 24 hours distant, coronas usually suggest imminent precipitation.

The flatness of these stratiform clouds is a consequence of what meteorologists call *stability.* Stability is simply the resistance of air to some force that is attempting to push it upward. Stable air tends to spread out in a layer; unstable air tends to balloon upward, like the bubbles in a pot of boiling water. When air cools very slowly with increasing altitude, or when warm air actually overlies cooler air near the surface, as in a warm front, the air mass is very stable. The result is often a thick, flat sheet of stratus clouds that may stretch from hundreds of miles offshore to the Cascades and Coast Mountains and possibly even farther east.

Warm-Front Clues

The general rules for gauging whether a warm front is approaching by observing clouds are:

- Look for approaching clouds, usually from the southwest, west, or northwest.
- Look for flat, sheetlike clouds (stratus).
- Look for thickening, lowering clouds.
- Look for surface winds from the east to southeast.
- Look for a decrease in air pressure.
- Look for an increase in air temperature.

Under these circumstances expect steady, widespread precipitation. (Mountain travelers must remember these are general guidelines and, depending on the orientation of the mountains, may not always hold true.)

Warm-Front Precipitation

There are no sure field rules for determining whether a warm front will produce snow or rain. For one thing, snow isn't limited to the winter months in the mountains of the Pacific Northwest. It can occur in any month of the year, depending upon elevation.

As a rule of thumb, assume that precipitation will remain as snow (even though it may not stick to the ground) down to approximately 1,000 feet (304 m) below the freezing level. Current and forecast freezing levels can be obtained from National Oceanic and Atmospheric Administration (NOAA) weather radio, avalanche hotlines maintained by the U.S. Forest Service, and in some of the print and broadcast media. We'll further discuss sources of weather information in Chapter 6, on pre-trip weather briefing.

It's important to remember that the freezing level reported is usually the free-air freezing level, that is, the altitude at which temperatures drop below freezing if uninfluenced by terrain effects such as trapped cold air in valleys or heating of the ground by sunshine.

Let's say, for example, that NOAA weather radio reports the freezing level at 4,000 feet (1,219 m). You're camped at 3,000 feet (914 m) on Oregon's Mount Hood, under a sky of dense gray clouds that have recently moved in. In this case, the precipitation is likely to begin as snow. As mentioned, this is an approximate rule that can be affected by a variety of factors. A locally heavy shower, for example, can lower the snow level to as much as 1,500 to 2,000 feet (457–609 m) below the free-air freezing level, because moderate to heavy precipitation can drag cold air farther down from the base of the cloud.

A second point is that precipitation tends to be spread unevenly. University of Washington researchers have determined that precipitation

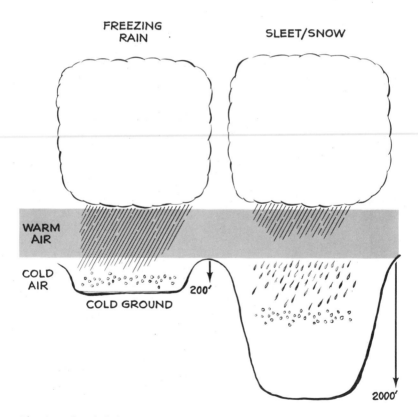

Fig. 15. *Sleet, left; freezing rain, right.*

tends to be clumped in cigar-shaped bands running parallel to a warm front. These bands tend to range in width from 5 to 20 miles (8–32 km). Therefore, the intensity of the precipitation will vary with time. Even a slow-moving, wet warm front will be marked by occasional decreases in precipitation, if not outright breaks.

Rain and snow aren't the only forms of precipitation produced by warm fronts, especially in the mountains. If the air beneath is below freezing, and the overlying air is sufficiently warm to produce rain, the warm front can produce either freezing rain or sleet (fig. 15). If the layer of cold air is shallow, freezing rain is most likely, whereas a thick cold-air layer is likely to produce sleet.

If, for example, you're snowshoeing in a valley with below-freezing temperatures, and the forecast freezing level is more than 2,000 feet (609 m) above your elevation, then freezing rain is likely to glaze over everything exposed to the elements. However, if the valley and all nearby passes are below freezing, the cold-air layer is thick enough that rain falling from the warm air aloft will probably freeze before it hits the ground. The result will be a good pelting with sleet or snow pellets.

Cold-Front Clouds

Perhaps after you've spent a morning trapped in a tent by precipitation from a passing warm front, the stratus clouds thin, possibly revealing streaks of blue sky. The air still feels relatively warm and moist. Then, before you break camp, the skies darken more than ever, and rain or wet snow falls more intensely than before. The brief interruption in precipitation was simply a sign that the surface warm front had passed; the new precipitation signals the arrival of a cold front (fig. 13).

Inland areas east of the Cascades and Coast Mountains often see more massive cloud buildups, a more definite break in precipitation, and some clearing (albeit hazy) between warm- and cold-front passage or, at times, only high cloudiness. But in the coastal regions, the transition from warm-front passage to cold-front arrival is usually brief, often marked by an accelerated drop in pressure and intensified precipitation.

In these transitions, warm, moist air no longer glides up and over the cooler air near the surface, so precipitation slows or ends. The air from the surface up is uniformly warm and moist. But then the cold air streaming down the back side of the approaching low collides with the warm, moist air, abruptly thrusting it upward. If warm fronts are the turtles of the meteorological world, moving slowly but steadily, then cold fronts are the jackrabbits.

Cumulus Clouds

The approaching cold air behind a cold front is much more dense, so the push upward can be very strong and fast, exceeding 20 miles per hour

(32 km/h). The result is not layered clouds but a line of towering heap clouds, called *cumulus clouds.*

When such clouds produce precipitation or thunder and lightning, we call them *cumulonimbus clouds.* These can extend as much as 50,000 feet (15,240 m) above the surface of the earth. However, in the Pacific Northwest, where approaching cold air has usually been warmed by its passage over the Pacific Ocean, the temperature contrast between the approaching cold air and the warm air already in place isn't as great, and cumulonimbus clouds don't grow quite as tall. Twenty thousand to 30,000 (6,096–9,144 m) is a more common maximum height in the Northwest, especially to the west of the Cascades and Coast Mountains.

◆

The exceptions to the rule of quick cold-front passage tend to occur when the cold front is sliding parallel to the coast, or mountain ranges, with south or southwesterly winds in the upper atmosphere. This situation is most common over the Coast, Olympic, and Vancouver Island ranges, and along the western slopes of the Cascade and Coast ranges.

Clues to the Approach of a Cold Front

- Look for brief clearing, or a decrease or end to precipitation.
- Look for clouds to thicken, lower, merge, and darken.
- Look for winds to increase, usually from the east or southeast, depending on mountain orientation.
- Look for a drop in pressure, usually rapid.

Under these circumstances, expect intensified precipitation with the front, colder temperatures after the front, and a wind shift to the southwest or west after the front, depending on mountain orientation.

Cold-Front Precipitation

As the University of Washington research team had found along warm fronts, precipitation ahead of and along cold fronts tends to be organized in cigar-shaped bands parallel to the front. This precipitation tends to be more intense because of the rapid upward movement of the moist air ahead of and along the cold front (fig. 16).

The upward movement is usually ten to a hundred times more rapid than that along a warm front. But whereas warm-front precipitation tends to be prolonged, often lasting as long as a day, cold-front precipitation is much more brief, due to the front's more rapid movement. Precipitation associated with a cold front generally lasts only an hour or two.

CLOUD COVER
AREA OF HEAVIEST
PRECIPITATION

Fig. 16. *Precipitation pattern from cold front, top view (courtesy Department of Atmospheric Sciences, Cloud Physics Group, University of Washington).*

Although the snow level usually extends only 1,000 feet (304 m) below the freezing level when stratus clouds are producing the precipitation, when the source is cumulus clouds, the snow level may extend as far as 2,000 feet (609 m) below the freezing level in heavy showers, especially after the cold front has passed.

Serious to severe thunderstorms are sometimes associated with the approach and arrival of cold fronts. This is true of the inland areas of the Northwest to the east of the Cascades and the Coast Mountains, where relatively warm, moist ocean air is in sharp contrast to the cold, dry air moving down from Alaska or the Yukon or Northwest Territories. But in the coastal regions, where the proximity of the Pacific Ocean moderates temperature contrasts, we tend to find thunderstorms after, not before or during, cold-front passage.

East of the Cascades and the Coast Mountains, clearing and drying usually follow the passage of a cold front. This isn't always the case for the western slopes of these ranges, or for the Coast, Olympic, and Vancouver Island ranges, which border the ocean. The areas that receive precipitation can be widespread or very localized, depending on the delicate interplay of terrain features and wind.

Occluded Fronts and Precipitation

An important variation on warm and cold fronts is the *occluded front* (fig. 17). Because of a sandwiching of cold and warm air, an occluded front combines the precipitation characteristics of both warm and cold fronts.

If you're experiencing prolonged rain- or snowfall with occasional

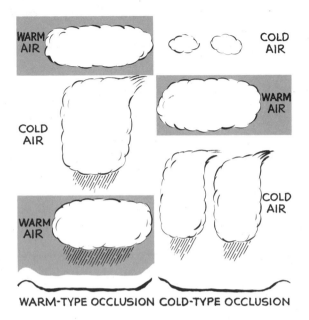

WARM-TYPE OCCLUSION   COLD-TYPE OCCLUSION

Fig. 17. *Precipitation pattern from occluded fronts: left, warm-type occlusion; right, cold-type occlusion.*

strong bursts, and possibly thunder and lightning too, an occluded front is the most likely culprit. Depending upon the direction of movement of the overall weather system, the passage of occluded fronts, like cold fronts, is much more rapid than that of warm fronts.

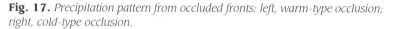

Fog

Fog, which is essentially a cloud at ground level, can cause many different problems for mountain travelers. It can make finding a route difficult if not impossible, obscure dangerous crevasses, and ice up rock at below-freezing temperatures.

Three types of fog are most prevalent in and around the mountains of the northwestern United States and British Columbia: radiation, advection, and warm frontal fog. Before exploring the subtleties of each, it makes sense to understand the fundamentals common to all.

The example of seeing your breath on a cold day is again useful here. To review: As we inhale air, it warms close to our body's normal temperature of 98.6 degrees Fahrenheit (37°C) and is moistened. As we exhale, that warm, moist air is rapidly cooled to the environmental temperature and may condense from water vapor (a gas) to water droplets (a liquid), just as steam from the shower condenses on the cold bathroom mirror. The result is a cloud at ground level, or fog.

Radiation Fog

Radiation fog is most common during the autumn months and is found both west and east of the Cascades and the Coast Mountains. It often follows wet weather.

Perhaps rain fell during your drive to the trailhead, followed by nighttime clearing, revealing a dazzling, star-filled sky. You are surprised to awaken to dense fog, possibly with light drizzle beading on the tent or on your forehead. But then, after breaking camp, as you climb to a peak overlooking the campsite, you notice the gray fog gradually lightening, the sun first appearing as a light disk then turning dazzling yellow as you emerge from the fog layer. In the Northwest, 1,000 feet down the trail or up the mountain can mean dramatically different weather.

Radiation fog is produced when damp ground (often saturated by precipitation) loses some of its moisture through evaporation to the lower layer of the atmosphere (fig. 18). As the disturbance that produced that moisture passes, the clouds clear. Overnight, considerable heat is lost, or radiated by the ground, to the colder atmosphere above. The moist air close to the wet ground cools and condenses into a layer of fog.

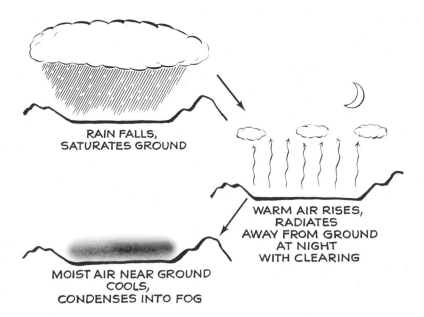

RAIN FALLS,
SATURATES GROUND

WARM AIR RISES,
RADIATES
AWAY FROM GROUND
AT NIGHT
WITH CLEARING

MOIST AIR NEAR GROUND
COOLS,
CONDENSES INTO FOG

Fig. 18. *Formation of radiation fog.*

41

Radiation fog is especially common in valleys and cirques, because cold air tends to drain downslope, hastening the cooling and condensation of the moist air below. It is occasionally rather thin, often only several hundred feet deep. Even when radiation fog is thick, it often burns off by midafternoon, then progressively earlier each day thereafter. However, if a trail is shaded by large peaks or even tall trees, the evaporation of the fog will be slowed considerably.

The same is true if the moist air is trapped close to the ground. This usually happens when warm air lies on top of colder air near the surface. Meteorologists call this situation an *inversion*, because air temperature normally cools with increasing altitude.

Radiation Fog Clues

- Moist ground from rain or melting snow.
- Clearing that allows extensive overnight cooling, especially in the late autumn to early spring, when nights are long.
- Light winds, generally 5 knots or less.

Advection Fog

While radiation fog is most common during the months when nights are long, *advection fog* (fig. 19) is most frequent during the summer months, primarily west of the Coast Mountains and the Cascades in British Columbia, Washington, and Oregon.

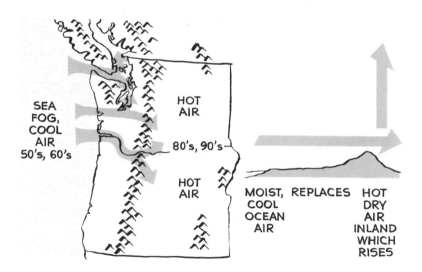

Fig. 19. *Advection fog.*

42

During this season, daytime air temperatures are considerably higher than those of the ocean water just off the coast, often by 30 degrees Fahrenheit (1.1°C) or more. The warm air rises, and cooler, more moist ocean air moves in to replace it. As the air temperature drops overnight, the moist ocean air condenses into fog or a low layer of flat stratus clouds. "Advection" refers to this horizontal movement of air from water to land.

Advection fog typically develops shortly before sunrise, surrounding the Coast, Olympic, and Vancouver Island ranges and blanketing the western slopes of the Cascades and the Coast Mountains to a maximum of roughly 5,000 or 6,000 feet (1,524/1,828 m) above sea level, but often less. Such stratus or fog rarely crosses the Cascades of Washington and Oregon or the Coast Mountains of British Columbia, so sun-seekers often choose destinations along the eastern slopes of these ranges.

Advection fog, like other weather phenomena, never moves in without warning, although the clues can be subtle. Here are the ones to watch for, in chronological order:

Advection Fog Clues

- Hot weather in Puget Sound/Willamette Valley and the Strait of Georgia/Queen Charlottes and/or eastern Washington, Oregon, and British Columbia.
- Cooling temperatures and fog moving northward along the coast.
- Wind shift in Puget Sound/Willamette Valley or Strait of Georgia/Queen Charlottes from north or northeast to south or southeast.
- Strong westerly winds in the Strait of Juan de Fuca, the Columbia River Gorge, or the Fraser River Canyon, or northwesterly winds through the Queen Charlotte Strait.

Marine advisories and warnings can be valuable aids in gauging how thick the fog or low-cloud layer is likely to be in the mountains, since such alerts reflect the strength of the push of ocean air into the interior. In particular, look at the strength of winds through the Strait of Juan de Fuca or the Columbia Gorge. Remember, wind speeds are given in knots (1 knot equals 1.15 miles per hour), and the wind direction must be from the west to drive the moist marine air inland.

Marine Advisories/Advection Fog Guidelines

Advisory	Fog/Stratus Depth	Precipitation	Burn Off
None (10–25 kn)	1,000–2,000 feet (304–609 m)	None	Afternoon
Small Craft (26–33 kn)	2,000–4,000 feet (609–1,219 m)	Drizzle?	Late P.M.
Gale Warning (34–47 kn)	4,000 feet + (1,219 m+)	Drizzle/rain	Next day

The thickness of the advection fog or stratus layer offers clues as to how soon it may burn off. If the layer is more than 2,000 feet (609 m) thick, it may not clear at all on the first day, perhaps not until midafternoon on the second day, and then progressively earlier after that. Incidentally, southwesterly winds off the coast tend to thicken fog layers; northwesterly winds tend to thin them.

Warm Frontal Fog

Our two previous types of fog offer the hiker, climber, skier, or snowshoer some options: move east of the Cascades or the Coast Mountains to escape advection fog; move higher to escape advection or radiation fog. The third type of fog offers no such options, outside of boarding a jet and heading for another part of the country. *Warm frontal fog* (fig. 20) covers a large area and is just as common east of the Cascades and the Coast Mountains as to the west.

Warm frontal fog is produced by precipitation falling from warm air aloft into colder air near the surface. The precipitation saturates the cold air, which then condenses into a thick layer of fog.

Unlike other varieties, warm frontal fog doesn't burn off. It disappears only after the surface warm front has arrived, bringing an end to the contrast between the cooler air near the ground and the warmer air running over it, and the precipitation formed by this process.

In the interiors of the United States and Canada, such fogs can persist for days during the winter. In the coastal regions west of the Cascades and the Coast Mountains, the duration is much shorter, simply because warm fronts tend to move through more quickly, and because the contrast between the air masses is not as great.

Warm Frontal Fog Clues

- Lowering, thickening stratus clouds.
- Light east to southeasterly winds.
- Steady precipitation.
- Small clouds with a shredded or torn appearance forming close to the ground.

Ice Fog

There's one final type of fog, much less prevalent than the three already discussed. That's *ice fog,* and it's chiefly found in the Fraser Plateau of British Columbia, between the Coast Mountains and the Rockies. It's also very common in the Yukon, the interior of Alaska, and the Northwest Territories. Composed of suspended ice crystals, ice fog is rare at temperatures warmer than –20 degrees Fahrenheit (–29°C). However, it becomes progressively more common with colder temperatures near a source of

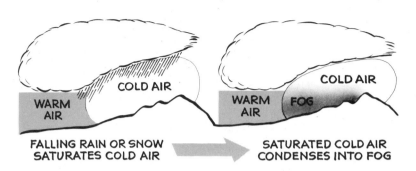

Fig. 20. *Warm frontal fog.*

water vapor, such as fast-flowing streams or large herds of animals. Ice fog is usually very localized but can be very dense.

The difference between the impact of fog and that of low clouds can be an important one in the mountains. Wind speed is usually the determining factor, as wind mixes some of the moist air near the ground with drier air aloft. That "lifts" the fog into a layer of flat, low clouds called stratus, or evaporates it altogether. Wind speeds of 5 knots or less are most conducive to fog formation; wind speeds greater than 5 knots usually lead to the formation of stratus clouds. However, moisture moving upslope can develop thick fog with wind speeds of up to 10 knots.

Look for such upslope fog to develop on the western slopes of the Cascades with a westerly wind, or the eastern slopes of the Cascades with an easterly wind, particularly during the winter after a thaw has melted snow, saturating the air with evaporated moisture.

CHAPTER 3

Localized Precipitation Patterns, or, A Mile Makes a World of Difference

◆

C *limbers and hikers frequently compare notes after a weekend in the mountains, often finding that the weather varied greatly over relatively short distances. A rock climber who spent the weekend at Smith Rock in eastern Oregon may have enjoyed sunshine and warm temperatures, while a friend who attempted to scale Mount Hood in the northern Oregon Cascades was driven back by rain, snow, and howling winds. Choosing destinations wisely requires an understanding of why and how weather varies over short distances in the Pacific Northwest.*

Postfrontal Precipitation

Postfrontal precipitation is simply the rain or snow that falls *after* a cold front moves through a region. It's a showcase for the extreme local variations in weather that occur in the Pacific Northwest. The colder air following the surface cold front streams over the Pacific Ocean, and two major changes occur: water evaporates into the cooler, drier air above; and the air is heated by the ocean, which may be 10 to 60 degrees Fahrenheit warmer than the air above. This is a prime example of unstable air.

To visualize what happens, think of a pot of water that's being heated on the stove. The water on the bottom is expanding through heating, small bubbles of air blossoming into bigger bubbles. Needing more room, this water is forced to ascend. Unstable air behaves the same way. Water vapor from the great moisture source of the Pacific Ocean cools and condenses into water droplets, which are carried aloft by the rising air, developing into towering cumulus or cumulonimbus clouds.

The upward nudge that generates these clouds may come from the air moving over the warmer ocean, or it may come from the air moving over the sun-warmed land of coastal Washington, Oregon, or British Columbia. Even partial breaks in the cloud cover can sufficiently warm the land to provide that boost.

And there's a third source to generate the upward shove: wind carrying the unstable air against the mountains. Because mountains act as barriers to moving air, when the two collide, the air has but one direction to go — up! As the air rises, there is frequently sufficient cooling to jump-start the growth of clouds and precipitation. The name for rain or snowfall produced in this way is *orographic precipitation*.

The lifting of air over mountains with even an average wind is often fifty to a hundred times as great as the lifting along a cold front. This creates tremendous amounts of precipitation on the windward side of ranges such as the Cascades and the Coast Mountains of British Columbia. In fact, roughly 60 percent of the precipitation falling on these ranges is orographic precipitation, created after, not during, frontal passage.

This can be seen in average annual precipitation amounts of areas at different elevations along the western slopes of the Cascades. Studies indicate that even in areas of small topographic variation, precipitation can vary as much as 30 percent for stations as little as 5 miles (8 km) apart. In fact, slight shifts in wind direction can radically change the locations of peak precipitation.

Washington

An excellent study of such precipitation patterns was conducted by Pamela Speers Hayes as part of her master's thesis at the University of Washington (1986). She has been kind enough to allow us to excerpt her findings here. Although they relate to western Washington, some general conclusions can be drawn that will be helpful in assessing precipitation elsewhere in the mountains of the Pacific Northwest. Hayes related the precipitation patterns to winds observed at approximately 5,000 feet (1,524 m) above sea level. We'll examine how to obtain such information in Chapter 6, dealing with weather briefings.

Southerly Winds

Southerly winds are associated with significant rain- or snowfall (fig. 21). Heaviest precipitation is found along the south side of the Olympics; at Mount Baker, Silverton, and Verlot, encompassing much of the North Cascades; and in the area extending from south of Mount St. Helens and southwest of Mount Adams to the Cascade foothills just north of the Columbia River.

Southwesterly Winds

Heavy precipitation is also common given southwesterly winds. The Olympics receive more precipitation than the Cascades; however, the focus shifts from the south side of the Olympics to near the Quinault ranger station. The North Cascades continue to be a wet zone, with the maximum extending farther east, from Mount Baker to Diablo Dam. The central Washington Cascades ranging from Skykomish to Snoqualmie Pass are another area of maximum precipitation.

47

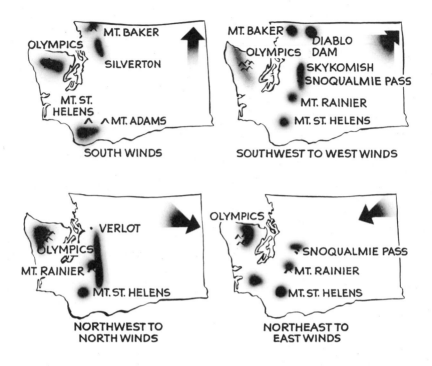

Fig. 21. *Wind direction/peak precipitation tendencies, Washington.*

Westerly Winds

Precipitation intensity generally decreases with westerly winds. However, when wind speeds exceed 25 miles per hour (40 km/h), the heaviest precipitation intensity of any wind direction is found with westerly winds. The key, then, is to check wind speeds as well as wind direction. The west side of the Olympics is one wet zone; the area from Skykomish south to Stampede Pass is another, with islands of maximum precipitation around Mount Baker, Mount Rainier, and Mount St. Helens.

Northwesterly Winds

With northwesterly winds the heaviest precipitation shifts from the Olympics to the Cascades, especially in the zone from Verlot south to Mount St. Helens. A "rain shadow" exists along the eastern and southern slopes of the Olympics; contrast this with the rain shadow north and east of the Olympics, when there are southwesterly winds. The rain shadow in the lee of the Olympics shifts as the wind direction changes.

Northerly Winds

Northerly winds are generally associated with fair weather, as they usually coincide with building high pressure and a drier offshore flow.

However, when precipitation does occur, it's generally in the form of localized showers, most commonly along the northwest side of the Olympics and around Mount Rainier, Snoqualmie Pass, and the zone from Verlot southeast to Stevens Pass in the Cascades.

Northeasterly Winds

Precipitation is very light when it occurs with northeasterly winds. When it does, most frequently in the winter months, it's found chiefly along the northeastern side of the Olympics, the Coast Range in southwestern Washington, Mount St. Helens and Mount Rainier, and Snoqualmie Pass.

Easterly Winds

Precipitation is rare and very light when associated with easterly winds, except in the winter months. It is most common in Snoqualmie Pass (as will be discussed shortly).

Southeasterly Winds

Significant precipitation is still rare when winds blow out of the southeast, although such winds commonly signal an approaching weather system, with heavier precipitation on the way. Significant precipitation is most common along the east and south sides of the Olympics, extending along the Coast Range into Oregon, and along the eastern slopes of the Cascades from Mount Baker to Mount St. Helens.

Although this data refers specifically to Washington, some important lessons can be learned for trips elsewhere in the mountains of the Pacific Northwest. Let's examine Oregon first.

Oregon

Oregon's Coast Range is a more or less continuous stretch of mountains extending from the California border to the Columbia River, to the west of and parallel to the Cascades. This lengthy range provides for impressive variations in precipitation given different wind directions.

South Through West Winds

South, southwest, and westerly winds produce the heaviest precipitation along the Coast Range, with peak amounts along the western slopes (fig. 22). Precipitation is also heavy along the western slopes of the Cascades from Mount Hood south to Mount Jefferson, decreasing farther south. In part, this is due to the rain shadow created by the more rugged terrain, especially where the Klamath Mountains merge with the Cascades and the Coast Range. However, the general storm track that brings weather systems to the Pacific Northwest is another factor; it most generally stretches from the northern half of Oregon into Washington and southern British Columbia. Lesser precipitation amounts are found along

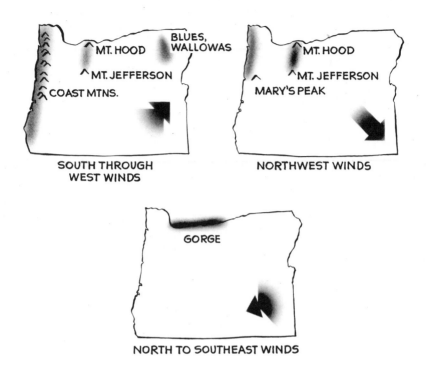

Fig. 22. *Wind direction/peak precipitation tendencies, Oregon.*

the southern and western slopes of the Blue Mountains in eastern Oregon. Precipitation there is generally half that found in the Cascades.

Northwesterly Winds

Precipitation with northwesterly winds is even lighter than that found in Washington, with the greatest amounts found along the Coast Range north of Marys Peak. The reason for this reduction is that as you move farther south, there's generally less of the cool, unstable air that produces rain or snow showers. Also, the jet stream, which directs storms, is more frequently found moving into Washington or British Columbia. When northwesterly winds do occur, the precipitation consists more of showers than of sustained rain or snow. If clouds are getting taller and merging together with a northwesterly flow, expect more frequent showers from Mount Jefferson north to Mount Hood.

North to Southeasterly Winds

North, northeasterly, easterly, and southeasterly winds rarely bring significant precipitation to Oregon. The major exception is during the winter, when an outbreak of cold air pushing from east to west through gaps or gorges can collide with warmer, moist air near the passes, producing locally heavy precipitation, usually snow but occasionally freezing rain. This is especially common along the Columbia Gorge, producing the well-known and generally detested "silver thaw." (We'll cover this phenomenon in greater detail later in this chapter.) When this happens, the western slopes actually offer the best weather conditions.

British Columbia

Southerly Winds

In British Columbia, southerly winds, commonly preceding a major weather system, lead to widespread precipitation, with maximum amounts noted along the southern and western slopes of the Vancouver Island Range. In the Coast Range, the heaviest precipitation tends to be found from Monarch Mountain northward. The Whistler area also experiences some of the heaviest precipitation.

Southwesterly to Westerly Winds

Southwesterly and westerly winds again tend to produce the heaviest precipitation along the western slopes of the Vancouver Island Range, and from Monarch Mountain northward, with a dry zone along the eastern slopes of the Vancouver Island Range and to the south of Monarch Mountain (fig. 23).

Northwesterly Winds

Northwesterly winds often bring showers that are heaviest over the northern and western ends of the Vancouver Island Range, particularly from Cape Scott to Golden Hinde. The Vancouver Island Range produces less of a rain shadow over the Coast Range, with significant precipitation from showers extending at least as far as Mount Gilbert. These variations are due to the added lifting of air along the windward slopes, which

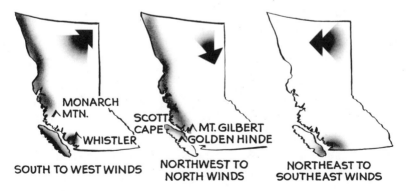

Fig. 23. *Wind direction/peak precipitation tendencies, British Columbia.*

51

wrings out additional moisture from the clouds, and the subsequent drying of air as it descends over the leeward slopes.

North to East Winds

In British Columbia, precipitation associated with north, northeasterly, or easterly winds is relatively rare; when it does occur, it is light. The exception, as mentioned in the discussion of Oregon, is during the winter, when cold arctic air flows through major river valleys or passes, colliding with and lifting the warmer, moisture-laden ocean air. This is especially common along the Fraser River Valley.

Thunderstorms and Lightning

Because the postfrontal precipitation associated with southwesterly to northwesterly winds develops in cumulonimbus clouds, it is also the greatest source of thunderstorms along the western slopes of the Cascades and coastal ranges. Pacific Northwest thunderstorms tend to be tame in comparison with their midwestern cousins. But consider that even an early spring thunderstorm ramming up against the western slopes of the Cascades can release up to 125 million gallons (over 473

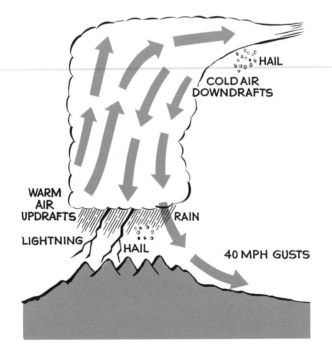

Fig. 24. *Thunderstorm hazards.*

million liters) of water. A blast of cold air from such a thunderstorm can produce winds gusting 40 to 80 miles per hour (64–128 km/h), posing a threat to climbers in exposed positions.

This heavy discharge of rain can quickly flood streambeds and small valleys; floods set off by thunderstorms have swept away entire campgrounds. But the greatest number of thunderstorm-related fatalities (outside those caused by tornadoes, which are rare in the mountains) comes from lightning (fig. 24).

The exact cause of lightning still eludes scientists. However, researchers know that the initial discharge stroke, or "leader," moves almost invisibly from the cloud toward the ground. No thicker than a pencil, this discharge stroke attracts electrical charges on the ground, which connect with the leader and race upward toward the cloud in the much larger return stroke, which is what we see.

A single lightning bolt may have an electrical potential of 125 million volts, heating up to 50,000 degrees Fahrenheit (27,732°C). That's five times the temperature of the sun's surface. Two hundred people die from lightning strikes in the United States each year, some in the mountains, including those of the Pacific Northwest. With a few precautions, most such accidents could be avoided.

If Thunderstorms Are Forecast

- DO NOT camp in a narrow valley or gulley.
- DO NOT plan to climb or hike in high, exposed areas.
- DO watch small cumulus clouds for strong, upward growth.
- DO keep track of weather reports.
- DO listen for strong static interference on AM radio broadcasts (these may come from lightning).

If You Spot Thunderstorms

- DO get inside a car or building if available.
- DO get away from water.
- DO seek low ground in open valleys or meadows.
- DO move immediately if your hair stands on end.
- DO NOT stand under trees, especially in open areas.
- DO NOT remain near or on rocky pinnacles or peaks.
- DO gauge the movement of thunderstorms.

As for that last point, how do you gauge the movement of thunderstorms? It's easy if you have a watch. The moment you see lightning, start counting the seconds. Stop timing once you hear thunder. Divide the number of seconds by five; the result is the distance of the thunderstorm from you in miles. Continue to time lightning and thunder discharges to judge whether the thunderstorm is approaching, remaining in one place, or receding. If the time interval between the lightning and thunder is decreasing, the thunderstorm is approaching; if the interval is increasing, it's moving away.

The above technique works because light moves from one place to another at very high speed — 186,000 miles per second (297,000 km/h), essentially instantaneous. The sound of thunder moves much slower, about a mile (1.6 km) every five seconds. If you don't hear thunder, the storm is probably at least 15 miles (24 km) away, as thunder can be difficult to hear at that distance.

This timing technique can be a big help, as I discovered late one night in the Canadian Rockies, just outside Banff, when I was awakened by thunder. The walls of my tent suddenly lit up like blue neon. Wide awake, I needed to make some decisions at a time when the inviting warmth of my sleeping bag sharply contrasted with the chill of the late-September night. My tent was pitched in a valley ringed by Mount Rundle and several other towering peaks, so we weren't dangerously exposed to direct lightning, but side flashes could still pose a risk. If the thunderstorm moved closer, I'd have to consider a hike back to the campground shelter. Using the technique just described, I started timing the intervals between the flashes of lightning and the following thunder. I found they were remaining the same, or lengthening. That meant the thunderstorms were keeping their distance, so after a half-hour of bleary-eyed observations, I snuggled back into my sleeping bag, enjoying the atmospheric night lights playing off the sides of the tent, and gradually dozing safely off to sleep.

———— Convergence Zone Precipitation ————

While thunderstorms in the Pacific Northwest mountains are frequently the result of cool, unstable air moving over the warmer water of the Pacific Ocean, or from rapid lifting against the mountains, there are other causes. One is the Puget Sound Convergence Zone.

The Puget Sound Convergence Zone (fig. 25) can produce locally heavy precipitation while other areas just a few miles to the north or south may be enjoying sunshine. The zone develops after the passage of a cold front, as high pressure builds into the coast, producing west to northwesterly winds.

When the onshore flow of air runs into the Olympics, it splits, some flowing through the Strait of Juan de Fuca to the north, some flowing through the Chehalis Gap to the south. The Cascades present an almost insurmountable barrier to the east, so some of the air moving through the strait is forced south into Puget Sound, while some of the air moving through the Chehalis Gap is forced north. The two opposing currents collide, forcing some of the air to rise and then be pushed into the Cascades by the winds passing over the Olympics.

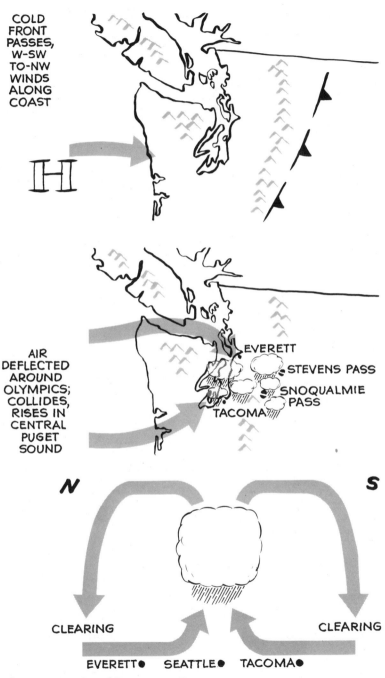

Fig. 25. *Puget Sound Convergence Zone.*

This convergence zone ranges from Everett to Tacoma, tending to move south during the afternoon and early evening hours. A weak convergence zone may produce only thicker clouds within central Puget Sound and against the central Washington Cascades; a stronger zone may produce locally heavy rain or snowfall, and possibly even thundershowers. Convergence zones have produced as much as 10 to 20 inches (25–51 cm) of snow at Stevens Pass, while only 2 to 3 inches (5.1–7.6 cm) of snow fell at Snoqualmie Pass, less than 50 miles (80 km) away. The reverse can also occur, depending upon where the convergence zone forms and how it moves.

Generally, in the Cascades, the convergence zone ranges between Stevens and Snoqualmie passes, although it may reach from Mount Pilchuck to Huckleberry Mountain.

Convergence Zone Clues

- Cold-front passage.
- Coastal west-southwest or northwesterly winds.
- North to northwesterly winds near Everett.
- Southerly winds to the south of Seattle.

Because the air that rises within the convergence zone sinks to the north and south, there occasionally is clearing in these areas, and certainly reduced precipitation. The best plan if a convergence zone is active or likely to be active is to head to the eastern slopes of the Cascades, or, if you're tied to the western slopes, plan your outing from Glacier Peak north, or from Mount Rainier south. Usually, the precipitation will be less intense and more showery.

Precipitation and the Time of Day

When moist ocean air dominates the Pacific Northwest after a cold front moves through, both time of day and location affect the likelihood of precipitation. Land, sea, and valley breezes are responsible for this pattern. We'll discuss how these breezes are generated in greater detail in Chapter 5, but first let's look at how they influence local precipitation.

Afternoon/Evening Hours

Because sea breezes that flow from water to land are most prevalent during the afternoon and early evening hours, they tend to thicken clouds that are up against surrounding mountains or hills (fig. 26). Examples include the western slopes of the Cascades in Washington and Oregon, the

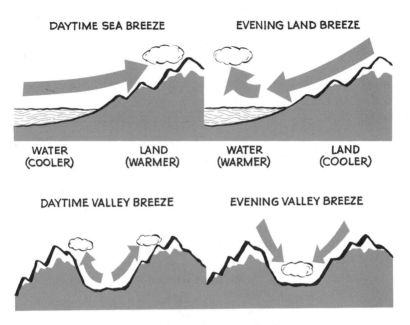

Fig. 26. *Breeze effects: top, sea breeze; bottom, valley breeze.*

eastern slopes of the Olympics in Washington, and the Vancouver Island Range. If the air is moist and unstable, its flow up these slopes can produce showers or even thundershowers. Therefore, the afternoon hours are the most likely time for showers or thundershowers along the western slopes of the Oregon Cascades, or the eastern slopes of the Coast Range along the Willamette Valley. The same is true for the western slopes of British Columbia's Coast Mountains or the eastern slopes of the Vancouver Island Range.

Late Night/Morning Hours

After sunset, the flow of air is downslope, toward Puget Sound in Washington, the Willamette Valley in Oregon, or the Strait of Georgia in British Columbia. This reduces the chance of precipitation along the mountains, and increases the chance of precipitation on or near the water or adjoining land.

Guidelines

If you suspect the air may be unstable, such as just after the passage of a cold front with relatively light and variable winds, you might time your hiking or climbing for the morning hours in such areas. That will help you avoid the typical afternoon showers or thundershowers. The same is true when moist ocean air blankets the region with low clouds, although drizzle or light rain is probably the worst that hikers or climbers will experience.

Eastern-Slope Precipitation

Perhaps the biggest source of moisture on the eastern slopes is produced when a low-pressure center develops or passes to the south of a range (fig. 27). This often occurs when, for example, a low-pressure center tracks up the Columbia River Gorge, or when a low-pressure center intensifies just to the southeast of the Cascades.

This pattern shifts the *orographic lifting,* or lifting of air caused by collision with mountains, to the east-facing slopes, producing much heavier

Fig. 27. *Eastern-slope precipitation.*

snow- or rainfall than normal. Hikers or climbers seeking drier weather will find it on the western slopes in this situation.

Many skiers or hikers are baffled when the western slopes of the mountains are bathed in sunshine, yet low clouds and snow envelope the mountain passes and the normally sunny eastern slopes. The transition from sunshine to snow may be abrupt, literally just over a ridgeline. This odd pattern is the result of the north-south orientation of the Cascades and other coastal ranges, which creates a barrier between the mild, moist ocean air west of the mountains and the drier air to the east, with its greater extremes of temperature (fig. 28).

During the winter months, a dome of cold air frequently develops east of the Cascades. This is caused by the land radiating its heat away under clear night skies, and snowpack reflecting much of the incoming solar radiation during the day. As the air within this dome of cold air cools, it occasionally reaches saturation point. Low clouds and fog form, which

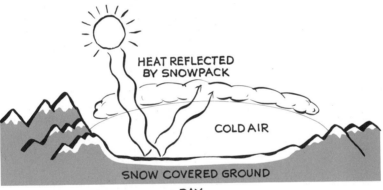

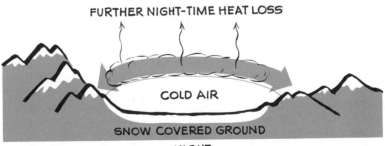

Fig. 28. *Inland winter cooling (cold dome).*

are usually very persistent thanks to the temperature inversion that normally develops as a result of the intense cooling of the air close to the ground. Snow flurries often begin, although not usually resulting in significant accumulations. The very cold air that has become saturated or nearly so has now primed the eastern slopes for major precipitation.

As the dome accumulates more cold air and grows higher, it pushes up against the eastern slopes of the mountains, which hold back the cold air from the warmer, less dense air west of the mountains, just like a dam holding back water. Frequently, the dome will reach the elevations of passes (fig. 29), such as Snoqualmie Pass in the Washington Cascades or Pengra Pass near Diamond Peak in the Oregon Cascades, or major river gorges that cut through the Cascades, such as the Fraser River in British Columbia and the Columbia River between Washington and Oregon. The passes or river gorges act as breaches in the dam that allow the cold air to seep through to the west side, often creating low-level clouds and precipitation.

This can be particularly dramatic during fair weather. The weather west of the Cascades may be clear and sunny with little wind. As the cold air pushes through the passes from the east, it can form a wedge of cold, cloudy, and even snowy weather that is often only a few thousand feet thick, or less. As you hike or snowshoe above a pass under these circumstances and you climb above the cold dome of air, you may be shocked to go from cold, snowy weather with brisk easterly winds to bright sunshine, mild temperatures, and light westerly winds. This transition occasionally takes place in only a few hundred vertical feet, as the top of the cold air is often sharply defined.

In extreme cases, the dome of cold air may build up higher than the Cascade crest; it then flows over the whole "dam." In this case, the entire crest and eastern slopes will be shrouded in cold, cloudy, and possibly snowy weather, while the west side of the Cascades and some of the higher peaks will still be in sunshine.

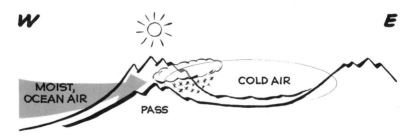

Fig. 29. *Localized snow generation in a pass.*

Guidelines for Anticipating Localized Snow and Low Clouds in Mountain Passes

Look for:

- Daytime temperatures well below freezing east of the mountains.
- Pass temperatures below freezing (you can obtain these by calling snow-line road reports).
- Freezing or snow levels forecast to rise 1,000 feet (304 m) or more above pass elevation.

Silver Thaw

As a dome of cold air flows from the east through the passes, it thrusts under the warm, moist air engulfing the region to the west of the Cascades. Occasionally, this warm air is producing rain, and as it falls into the shallow layer of cold air below, that rain freezes, glazing over anything and everything.

Such a pattern is often called a "silver thaw," because the coating of ice gives trees, rocks, and roads a silvery white appearance. Such patterns are most common along the Fraser and Columbia rivers during the months of December, January, and February. The worst cases occur when the Pacific Northwest has been locked in subfreezing weather, and a low-pressure system is approaching from the west, bringing warmer air and the promise of rain.

The "Back-Door Cold Front"

A silver thaw can persist for several days, but when it breaks down, even more interesting weather can develop. As the low-pressure center approaches the coast, pressures west of the Cascades begin to fall. This pulls cold air through the passes even harder, increasing the easterly winds and producing more lifting and more precipitation.

As a warm front approaches the Cascades, the warm, moist air is forced up and over the cold air in the passes. This lifting of warm air creates clouds and precipitation on the west side and through the passes. Often, the air moving in from the west raises snow levels well above the top of the cold air near the surface. The result? Sleet or freezing rain, and a heavy demand for tow trucks.

After the warm front moves through the passes from the west, it's followed by a cold front moving in the same direction; this air is colder than the warm air just ahead. Pressures fall rapidly east of the Cascades. The winds in the passes shift from east to west, destroying the temperature inversion. But because the cold air seeping through the passes from the east is frequently colder than the air behind the cold front moving in from the Pacific to the west, the temperatures at the pass actually rise after the cold front's passage!

Back-door cold fronts are most common during the winter months in the Northwest and therefore serve as a significant source of snow for eastern Washington, Oregon, and British Columbia. (Other effects caused by the cold air east of the ranges will be discussed in Chapter 5, on winds). Because the cold dome of high-pressure air lifts moist air as it surges against the mountains, the precipitation (usually snow) occurs as the air pressure rises instead of falls.

How can you anticipate this? Compare current temperatures and humidities in the region with the forecast temperatures. If considerable cooling is forecast, and current relative humidities are high — greater than 60 or 70 percent — precipitation is possible. This is another case in which you escape the clouds and precipitation by moving to the western slopes of the mountains. It's another example of the unique weather patterns found in the mountains of the Pacific Northwest.

CHAPTER 4

Snow and Avalanche Conditions: How to Find the Best and Avoid the Worst

◆

*S*now carries with it special pleasures and special risks in the Pacific Northwest. As much as skiers and snowshoers vilify the wet, heavy snow commonly called "Cascade concrete," the first snowfall triggers a rush of adrenaline and an urge to drop everything and head for the mountains. While wet snow is common, it's possible to use an understanding of the atmosphere to find the best snow available.

Snow also brings the danger of avalanches, and the need to learn how to avoid them. However, even considerable knowledge is no guarantee of safety without vigilance. As André Roche, the founder of the world-renowned Swiss Avalanche Center, has said, "When you're out there, be careful — the avalanche doesn't know you're an expert."

During the winter of 1988–89, the head of the Mount Hood Meadows professional ski patrol was killed in a slide. He was at the bottom of a small slope, no more than 200 yards (183 m) wide with a vertical drop of perhaps 60 feet (18 m), when a snow slab released, burying him under 6 feet (2 m) of snow. Although dramatic slides receive most of the attention, small slides are in fact the biggest killers. Eighty percent of avalanche deaths occur in slides that travel less than 100 yards (91 m).

The point of this discussion isn't to keep mountaineers out of the mountains, but to instill an ever-present caution. An in-depth review is beyond the scope of this book, but the following is a summary of key safety principles.

Two excellent guides exist for further study: *The ABC of Avalanche Safety,* published by The Mountaineers Books, and the *Avalanche Handbook,* published by the U.S. Forest Service as Agriculture Publication 581.

Anyone who plans to venture beyond the confines of developed ski areas should take an avalanche safety course. The National Ski Patrol offers one through the American Red Cross, and many private courses exist as well. Investigate, and take one.

Finding Good Snow

Snow is fascinating stuff. It forms very differently from rain, yet what begins as snow often ends as rain. Liquid precipitation, or rain, is the result of collisions between tiny water droplets in a cloud. Given the right circumstances, the droplets stick to each other, eventually growing large enough to fall as rain.

If the cloud temperature is well below freezing, it may produce snowflakes. But air temperatures below freezing don't necessarily guarantee snow. Cloud droplets don't automatically freeze at 32 degrees Fahrenheit (0°C), but can persist as liquid at temperatures as cold as –38 degrees (–39°C). We call such water droplets "supercooled" droplets.

Once ice crystals do form in the frigid air within a cloud, aided by the presence of small particles of clay or other substances, the transformation of the supercooled water droplets into snowflakes proceeds like a row of dominoes toppling each other. Water evaporates from the droplets and is sublimated onto the growing ice crystals. This process typically continues until most of the cloud's droplets have evaporated and the cloud is a large collection of ice crystals. However, that's still no guarantee of snow on the ground and an early ski season.

If the air below the cloud is warm, the ice crystals will melt, falling as rain. But if that air is cold, and the ice crystals have grown large enough, they may fall out of the cloud as snow, in a variety of shapes so immense that it has been said no two snowflakes are exactly alike. (That claim has been challenged, and in any case, snowflakes can be grouped in families sharing similar shapes; see fig. 30.) The difference in the shapes of snow-

TEMPERATURE	ICE CRYSTAL TYPE	
25–32°F	THIN HEXAGONAL PLATE	
21–25°F	NEEDLES	
14–23°F	HOLLOW COLUMNS	
10–14°F	SECTOR PLATES	
3–10°F	DENDRITES	
–8–3°F	SECTOR PLATES	
–58––8°F	HOLLOW COLUMNS	

Fig. 30. *Types of ice crystals and formation temperatures.*

flakes is the result of the different temperatures at which they form.

The proximity of the Pacific Ocean and the fact that most weather systems moving into the Pacific Northwest originate over the Pacific mean that in this area snow has a higher water content than in mountain ranges farther east, such as the Rockies.

If you were to fill a box 1 cubic meter in volume (a little more than a cubic yard) with fresh powder snow from the Colorado Rockies and let it melt, the water would only weigh 60 to 70 kilograms, equivalent to 130 to 140 pounds. The wetter snow found in the Cascades and Coast Mountains would yield about 100 kilograms of water, or 220 pounds. Cascade concrete, the really wet, heavy stuff, might tip the scale at 140 kilograms, or 300 pounds, but believe it or not, it wouldn't come close to the new snow with the highest average measured water content in the world: with a recorded water content of 260 kilograms per cubic meter, or 570 pounds! That can be found just to the west of the Cascades, on the Blue Glacier in the Olympics.

Few climbers, snowshoers, or skiers would like to slog through that kind of ready-mix. Most skiers prefer powder, which lets them leave a rooster-tail plume to mark their turns. From late October to March, the best conditions during an active snowfall can be found where the air temperature ranges from 20 to 25 degrees Fahrenheit(–6.7° to –3.9°C). Jacket zipper pulls with thermometers on them can be helpful for judging snow conditions, as long as they're on the outside of the jacket (away from body heat).

Generally speaking, ski at or above the elevation of the freezing level (which you can obtain from the sources we'll discuss in Chapter 6). The best snow is usually 1,000 to 2,000 feet (304–609 m) above the freezing level. In the Pacific Northwest, that usually means elevations from 4,000 to 7,000 feet (1,219–2,133 m).

If you can't find out the freezing level but have a thermometer and an altimeter, you can make an estimate by using the temperature at your elevation and the "standard lapse rate," the average between what are called the moist and dry lapse rates, the rates at which air cools with increasing altitude above the earth's surface in saturated and unsaturated air. The standard lapse rate is 3.5 degrees Fahrenheit per 1,000 feet of elevation (or 2 degrees Celsius per 304 meters).

Here's an example, figured in feet and degrees Fahrenheit. Camped at 3,000 feet, you observe the air temperature to be 38 degrees. Subtract 32 (the freezing temperature) and multiply by 1,000 (feet). The result is 6,000. Divide this by the standard lapse rate of 3.5 degrees. The answer is 1,700, indicating that the freezing level should be 1,700 feet above the campsite. Add that to your current 3,000-foot elevation to get the estimated freezing level: 4,700 feet.

In other words, to estimate the freezing level, perform this simple calculation:

$$\text{YOUR ELEVATION} + \frac{(\text{CAMP TEMPERATURE} - 32°) \times 1{,}000 \text{ FT.}}{3.5° \text{ (STANDARD LAPSE RATE)}} = \begin{array}{c}\text{ESTIMATED} \\ \text{FREEZING} \\ \text{LEVEL IN FEET}\end{array}$$

$$3{,}000 \text{ FT.} + \frac{38° - 32° = 6 \times 1{,}000 \text{ FT.}}{3.5} = 4{,}714 \text{ FT.}$$

The freezing level is at 4,714 feet, approximately 1,714 feet higher than your campsite elevation. Here's the same example, figured in meters and degrees Celsius:

Camped at 910 meters (approximately 3,000 feet), you observe the air temperature to be 3.3 degrees. There's no need to subtract the Celsius freezing temperature, because that is 0. Multiply by 304 meters. The result is 1,064. Divide by the standard lapse rate of 2 degrees Celsius. The answer is 502 meters above the campsite. Add 502 to your existing elevation of 910 meters to get the estimated freezing level of 1,412 meters.

$$\text{YOUR ELEVATION} + \frac{\text{CAMP TEMPERATURE} \times 304 \text{ METERS}}{2° \text{ (STANDARD LAPSE RATE)}} = \begin{array}{c}\text{ESTIMATED} \\ \text{FREEZING} \\ \text{LEVEL} \\ \text{IN METERS}\end{array}$$

$$910 \text{ M} + \frac{3.3° \times 304 \text{ M}}{2} = 1{,}412 \text{ M}$$

Now, a word of caution. Temperature inversions are common in the mountains (that is, the air temperature decreases more slowly than the standard lapse rate or actually increases with altitude). Therefore, the standard lapse rate may not reflect reality; use it as an approximation. If winds have been light, and skies clear or covered by stratus clouds, odds are an inversion exists. If winds have been brisk or a weather system is approaching, present, or departing, an inversion is less likely, and the result from the formula for estimating the freezing level will be more accurate.

In spring or summer, the best strategy to maximize enjoyment on the snow is to start early. Plan to arrive at your highest elevation by the mid- to late-morning hours, and then head down.

Wind has a major effect on the quality of snow. When winds are light, ski high; when they're strong, ski or snowshoe below the timberline. If winds have been blowing out of the southwest, which is common, head for northeastern slopes. Snow on the northern slopes tends to last longer, so an excellent strategy for skiing or snowshoeing is to start out early on the southern slopes, and save the northern slopes for later in the day.

The storm track in the Pacific Northwest that's most likely to bring optimal snow conditions is out of the west-northwest. Gradually cooling temperatures can be expected, and the snow that falls will have increasing density with depth, yielding light snow on the surface but heavier snow below, giving good support for your skis so they don't bury themselves as easily. When the forecast indicates west to northwesterly winds aloft and an approaching storm, get ready to head for the slopes! The new snow should be superb.

Avoiding Avalanches

Just as fresh snow offers wonderful conditions in which to play, it also increases the avalanche hazard. To assess the avalanche hazard in the mountains, answer two questions: What weather conditions preceded this trip? and What is the nature of the slope?

From the Cascades westward, direct-action avalanches are most common, generally occurring during or within 24 hours of major precipitation. A direct-action avalanche is one that takes place during or just after snowfall. The weight of the new snow can trigger the avalanche, as can the added weight from a skier or vibrations from someone crossing the snowfield.

In a continental climate, such as in the Rockies, deep avalanches, which involve more than just sloughing of surface snow, can occur well beyond the 24-hour period. The eastern slopes of the Cascades and the Coast Mountains offer a hybrid of direct and deep avalanches; generally, avalanches will occur within 24 hours of major precipitation, but not always.

If a storm starts at cold temperatures and then warms, the avalanche risk will be greater. Wetter, heavier snow will rest on top of less-dense snow, an unstable situation that is essentially an avalanche waiting to happen. If unseasonably cold temperatures have dominated the region and a storm is approaching from the southwest or south-southwest, our so-called Pineapple Express storm track, exercise extreme caution. A good rule of thumb is to watch out when freezing levels are forecast to rise.

Winds of 10 to 15 miles per hour (16–24 km/h) or more add to the risk. Such winds tend to break up ice crystals, allowing the snow to pack together more efficiently in a slab (a compressed layer of snow), especially on lee slopes.

After reviewing weather conditions, assess the slope you plan to cross. An avalanche risk exists anytime objects that tend to anchor snow become buried. Examples include closely spaced boulders, old tree trunks, and small trees or shrubs. The frequency of avalanches is greatest on slopes ranging between 25 and 45 degrees. But don't relax your guard if you are crossing below a steep slope; such areas are called "runout zones" for good reason. A slab avalanche barreling down a slope doesn't magically stop at the bottom. Its inertia will continue the slide for some distance.

The runout zone will offer clues. In the Cascades and the Coast Mountains, in fairly dense timber, look for what's called the "trim pattern." If the lower limbs are missing on the uphill side, you're still within the runout zone, and at risk. A conservative guideline is that if you can ski through the trees you can have an avalanche, but if you have to hike through them an avalanche is very unlikely.

Although the Pacific Northwest experiences its heaviest precipitation from November through February, midwinter avalanche fatalities are rare. The number of backcountry travelers killed by avalanches has actually decreased, which is especially significant because more men and women are skiing, climbing, and snowshoeing in the backcountry. Excellent work by the U.S. Forest Service avalanche forecasters must be credited for much of the decrease in fatalities during the winter months.

The greatest number of avalanche deaths occur in the spring, particularly in late March and early April. The spring increase is attributed to what avalanche forecaster Rich Marriott calls the "bluebird syndrome," the belief that if skies are blue, nothing can happen to you. Marriott also believes the increase in fatalities at that time is partially due to hikers and climbers trying to get a jump on the season, when they're not familiar with snow conditions.

Snow conditions change in the spring. Longer days lead to increased melting. The water percolates down through the snowpack, and when it hits a layer of crust, it spreads out, acting as a lubricant for the slab above.

The trigger for such spring avalanches comes from the daily cycle of melting and freezing. When clouds develop, leading to warmer nighttime temperatures, the crust doesn't redevelop on the surface of the snowpack. The next day, the sun won't have to melt off the surface crust, the density of the layer will increase with increased snowmelt, and backcountry travelers should be prepared for a larger number of slides, especially from the late morning to early afternoon hours.

An example of why careful attention is so important comes from the experience of two avalanche experts. Late one warm, sunny April day these two were playing a game of Frisbee on skis. No avalanches had occurred, and the pair didn't want to lose any altitude, so they selected a long traverse across a relatively steep slope. Focusing on their Frisbee, they were surprised to suddenly hear "thunder," and were chilled to see an avalanche throwing up chunks of snow the size of compact cars, roaring down a section of the slope they had crossed less than ten minutes before. As André Roche says, the avalanche doesn't know you're an expert. Knowledge is no substitute for attention.

One of the most dangerous avalanche areas is the North Cascades of Washington. Rugged terrain surrounds narrow valleys, allowing avalanches to slide down one slope and continue up the next side. Anyone planning an outing in that area should be particularly cautious and willing to switch plans if the avalanche hazard is likely to increase.

After assessing weather and slope conditions, take the shortest path possible across an avalanche-prone area. If in a group, never expose more than one person at a time. While that one person crosses, everyone else should be watching carefully. If a slide occurs, speed is critical in recovering the victim; the odds of surviving an avalanche drop below 50 percent after just half an hour.

Anyone traveling in avalanche country should have as minimum equipment common-sense knowledge, a metal shovel, and an avalanche transceiver.

The common-sense knowledge comes not from reading this book or the other references mentioned, but from taking a course on the subject. Each year *The Avalanche Review* magazine's fall issue contains a list of courses and contacts. You can request a copy from Knox Williams at the American Association of Avalanche Professionals, in care of the U.S. Forest Service, 240 West Prospect Street, Fort Collins, CO 80526.

The metal shovel can be purchased from any well-stocked recreational equipment store. It's used for digging snow pits to assess avalanche hazard, and for digging out avalanche victims.

The avalanche transceiver may seem expensive, but in reality it offers cheap insurance. However, even the best transceiver is worthless unless you know how to use it and practice frequently. Each member of the party must have one.

Chapter 6 outlines how to get good weather information before heading into the mountains, and how to integrate that information. Before pressing on, consider one final thought: Most avalanche victims trigger the avalanches that kill them, so be careful.

CHAPTER 5

Mountain Winds

◆

*S*taking a tent when skies are clear and winds are light seems a small *and simple matter. It's a different situation in· high wind. Although tents aren't designed to fly, given the proper wind direction and speed, they can put some kites to shame.*

Understanding wind and the way it behaves in the mountains can at the very least save labor and embarrassment. It can also save your life. On high peaks, wind velocities can exceed 100 miles per hour (160 km/h). The force is great enough to rip climbers right off the mountain.

Wind — The Big Picture

Wind is nothing more than moving air, which tends to go from regions of high pressure to regions of low pressure. The greater the difference in pressure between two locations, the greater the pressure *gradient* and the faster the air will tend to move.

Winds in the Upper Atmosphere

Wind moves differently in the upper reaches of the atmosphere than it does at or near ground level (fig. 31). In the upper atmosphere, the tendency to go from high to low pressure, the pressure gradient force, is balanced by the Coriolis force, which results from the rotation of the earth.

The Coriolis force deflects the wind to the right of its original path. Thus, instead of flowing from high to low pressure, wind tends to snake between the major high- and low-pressure systems in the upper atmosphere, undulating from west to east in the Pacific Northwest.

When these winds exceed 50 knots, meteorologists refer to them as the jet stream. Below that speed, they're generally called "steering currents," because they do just that: steer the weather disturbances at the

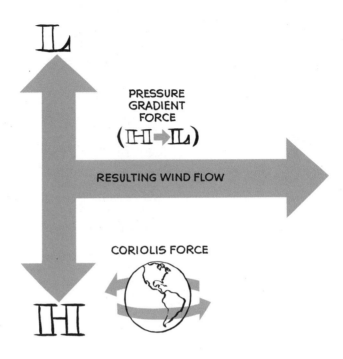

Fig. 31. *Wind flow in the upper atmosphere.*

surface. Different storm tracks bring different precipitation patterns (as discussed in Chapter 1 and reviewed in the final chapter, on field forecasting techniques).

Winds in the upper atmosphere strengthen when the contrast between temperatures in the midlatitudes (from 30 to 60 degrees) and those in the Arctic reaches a maximum. That's usually in the late autumn to early spring months. The Arctic, thanks to minimal heating from the sun, plunges into the deep freeze. The stronger the jet stream or steering currents, the stronger the weather disturbances at the surface of the earth, and the more likely they are to intensify.

Think of surface weather systems as riding on the steering currents of the atmosphere, somewhat like a boat on a moving river. For the purpose of this book, we can consider upper atmospheric winds to be those at and above 18,000 feet (5,486 m — above the elevation of Mount Rainier). Some television weather presentations show, at least in a general sense, the direction of the winds aloft.

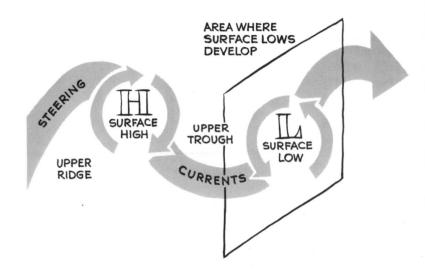

Fig. 32. *Relationship between surface systems and winds aloft.*

Surface high-pressure systems tend to be downwind of an upper ridge of high pressure, and surface low-pressure systems tend to be downwind of an upper trough of low pressure (fig. 32). Even if an upper trough hasn't produced a surface low-pressure system, one can develop in the area enclosed by the box. Seeing such an upper trough just offshore on weather maps should lead to caution and restraint in selecting a recreation site.

Fast winds high up in the atmosphere tend to produce fast winds at the surface. In the mountains, when formal information is scarce, watch the cirrus clouds that arrive in advance of an approaching weather disturbance. If you can see those clouds moving, the winds aloft are probably blowing in excess of 100 miles per hour (160 km/h). Expect strong, gusty winds in and near the mountains soon.

The higher the elevation you plan to reach, the stronger the winds you'll find. At the top of peaks such as Mount Hood and Mount Rainier, winds can easily exceed 100 miles per hour. Winds of 150 miles per hour (240 km/h) are not unheard of at such elevations.

Climbers, hikers, skiers, and snowshoers can obtain information on the winds aloft before setting out; we'll discuss several good sources in the next chapter. However, lacking such information, doubling the surface wind speed at sea level as observed over open water or land can at least offer a workable estimate of wind speeds between 5,000 and 10,000 feet (1,524 and 3,048 m).

Surface Winds

The direction of surface winds will be very different from the direction of winds aloft. Near the surface, the movement of air is somewhat more complicated than high up in the atmosphere. The pressure gradient and Coriolis forces are still present, but now friction comes into play.

The movement of air over water or land isn't smooth; the rougher the surface, the more turbulent the air flow over it, and the greater the friction. This upsets the balance between the pressure gradient and Coriolis forces. Air no longer flows parallel to high- and low-pressure systems; it now moves from high to low pressure, traveling clockwise out of and around a surface high and counterclockwise around and into a surface low. This tendency is reflected in a simple rule for locating high- and low-pressure systems when still in open terrain: *If you stand with your back to the wind, low pressure will be to your left.* Fig. 33 shows why this works.

Once the relative direction of a low is determined using this technique, a compass can be used to find the magnetic direction. Knowing that steering currents move surface weather systems from west to east, if the low is generally to the west of your position, it is likely moving toward you. If the low is to the east of you, it is usually departing.

This technique obviously works in the lowlands and on peaks, but it doesn't work quite as well on the lower slopes. Variation in wind direction caused by friction at elevations near or below surrounding terrain makes this guideline unreliable. Use the technique only on exposed positions at elevations higher than surrounding terrain.

Just as the direction of winds aloft plays a major role in determining

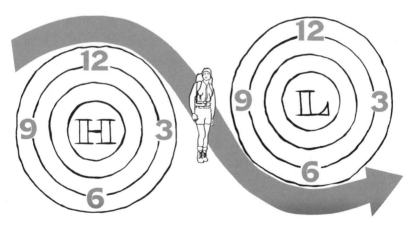

Fig. 33. *Finding a surface low.*

our weather, so does the direction of winds at the surface. It also provides valuable indicators of an approaching storm.

Pacific Northwest Mountain Wind Indications

- Surface winds from the north or northeast: fair weather likely.
- Surface winds switching to the east or southeast: a weather disturbance is probably approaching. Expect southwest to west winds with the passage of the front.
- Surface winds switching from east or southeast to southwest or west: expect brief clearing, but be prepared for the possibility of showers or thundershowers on western slopes.

It's not necessary to drag around a weather radio or a laptop computer capable of receiving weather telemetry to determine wind direction in the mountains. However, some weather radios are lightweight and compact, and if you will be skiing, hiking, or climbing at higher elevations where radio reception might be good, it's an option to consider.

The best and easiest way to keep track of such changes is to use your compass. As you orient yourself along your route, note the direction of the wind. Make a mental note, or jot it down in a small notebook or on the margin of your topographical map along with the time of your observation.

Changes in wind direction may be due simply to variations in terrain, or they may be due to changes in weather systems. Look for other confirmations, such as changes in wind speed. An anemometer that precisely measures wind speed isn't necessary; just be alert to your environment. For example, are the leaves or boughs of trees being rustled more forcefully along a wooded trail, or is more dust or snow being picked up along a trail in the open? An increase in the ripples or waves on a lake or stream may be another sign.

Changes in pressure lead to changes in wind speed as well as direction. Careful attention to air-pressure changes can provide early warning of high winds. An altimeter, which is frequently used as a route-finding and navigation tool by mountaineers, is essentially nothing more than a specially calibrated barometer. Watch for a continuous drop in air pressure registered on the altimeter as a continuous increase in altitude, despite the fact that you are walking along a level trail or actually descending. The altimeter registers the drop in pressure as a gain in altitude, though no such gain has taken place. Such false readings should serve as weather alerts.

Air-pressure decreases (altimeter increases) can be important in gauging possible approaching bad weather. Here are some guidelines that have proven useful in my experience both in the forecast office and in the mountains.

Pressure/Altimeter Indications over Three Hours

Pressure Decrease	Altimeter Increase	Recommended Action
.02 to .04 in. .6 to 1.2 mb (millibars)	20 to 40 ft. 6 to 12 m	None except for normal monitoring of sky.
.04 to .06 in. 1.2 to 1.8 mb	40 to 60 ft. 12 to 18 m	Watch sky carefully for thickening, lowering clouds. Wind increasing, shifting to east or southeast?
.06 to .08 in. 1.8 to 2.4 mb	60 to 80 ft. 18 to 24 m	Watch sky, wind carefully as above. Consider terminating the outing, due to the possibility of high winds, or seek safer alternatives.
.08 in. or more 2.4 mb or more	80 ft. or more 24 m or more	Terminate the outing, or seek the safest alternative available.

Wind — The Local Scene

Large-scale wind patterns are important, both at the surface and in the upper atmosphere. But by their very nature, the mountains of the Pacific Northwest alter wind considerably. Often, it's channeled through gaps in the terrain such as major passes, or even between two peaks. Wind speeds can easily double moving through such a gap.

Three of the best examples in the region (or perhaps the worst) are Snoqualmie, Stampede, and Naches passes in Washington's Cascades. These serve as the lowest passes north of the Columbia River. The most extreme conditions occur during the winter months, when temperatures east of the Cascades are far colder than those to the west, occasionally by as much as 50 degrees Fahrenheit.

Because the air to the east is more dense and therefore at higher pressure than the air to the west, it pushes against the Cascades, often eventually reaching the elevation of Naches and/or Stampede pass. It then surges through the gap, accelerating through the Green River Valley as it races toward Enumclaw, Black Diamond, and Maple Valley. Winds that are a strong 50 miles per hour (80 km/h) in the vicinity of Stampede

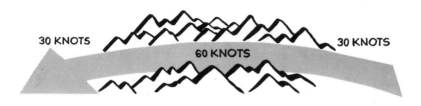

Fig. 34. *Wind acceleration through gaps and passes.*

Pass can easily exceed 100 miles per hour (160 km/h) as they move through the Green River Valley (fig. 34). Anticipate similar dangerous conditions in the Columbia River Valley between Washington and Oregon, and in the Fraser River Canyon in British Columbia.

The key lesson to be learned is to carefully examine surface winds upwind of a gap or pass and, in the case of a trip planned in the vicinity of such terrain features, to expect wind speeds to possibly be double those upwind.

Extreme blasts of wind occur when the pressure difference between, for example, Yakima and Sea-Tac (a distance of 104 miles [166 km]) increases to approximately 10 or 11 millibars (.30 inch of mercury) or more. Certainly, when the pressure difference from the upwind to the downwind side of a gap reaches or exceeds 4 to 5 millibars (.12 to .15 inch mercury per 100 miles), consider seeking immediate shelter from an exposed position.

If you listen to NOAA or Canadian weather radio before setting out on a trip (and I strongly recommend it), noting the different pressures reported by the observing stations can give you important clues that strong winds will develop in the mountains, especially in the passes.

Pressure Difference	Between	Winds 30 knots +
4 mb or .12 in. of mercury	Quillayute and Bellingham or Vancouver	In east-west passes and valleys
"	Astoria and The Dalles	In east-west passes and valleys
"	Astoria and The Dalles	In north-south passes and valleys

Terrain Blocking

Just as mountains can channel and accelerate wind, they can also provide protection from it. Obviously, climbing, skiing, or hiking on the downwind or leeward side of a peak will offer considerable protection from the wind. However, a major peak will also offer some limited protection to other peaks just downwind. A little knowledge of terrain blocking may provide more comfortable options.

The flow of wind around a cone-shaped peak such as Mount Rainier will be split for some distance downwind. That separation zone (fig. 35) generally extends to five times the height of the mountain, as measured from the summit of the mountain downwind (fig. 35). Given Mount Rainier's height of 14,411 feet (4,392.4 m), then, the air flow around the mountain would be divided for a horizontal distance of 72,050 feet, or approximately 13.5 miles (21.6 km). To look for alternate peaks, measure that distance from the summit, not the base of the mountain, on the leeward side.

Use this guideline conservatively. Gusty winds or nonconical peak shapes shrink the separation zone. An excellent technique to aid in visualizing the flow of wind around mountains is to slide your topographical relief map into a waterproof cover, and use a grease pencil or water-soluble marker to draw arrows depicting wind directions reported or forecast before departure, or the wind directions observed on wind-exposed sites. This practice will help you find out what areas are likely to be sheltered from the wind, how wind will tend to follow the shape of the land, and where gaps will tend to accelerate wind.

Valley Winds and Gravity Winds

Differences in the heating of bare ground or rock, as opposed to ground covered by vegetation or trees, can produce *mountain* or *valley winds.* As the ground heats during the day, the air close to it also heats and rises, moving up either side of a valley and spilling over the adjoining ridge tops. Such uphill breezes can reach 10 to 15 miles per hour

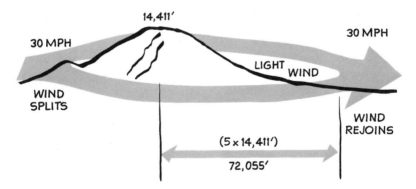

Fig. 35. *Terrain blocking of wind.*

(16–24 km/h), attaining peak speed during the early afternoon hours and dying out about a half hour before sunset.

At night the land cools, and the cool air flows downslope in what is called a *gravity wind*. Such downslope breezes reach their maximum after midnight, dying out just before sunrise. Speeds are similar to those of the uphill breezes, but can easily be stronger at the base of a snowfield or steep slope or cliff.

Camping at the base of a cliff, then, may result in an uncomfortably breezy evening. The steeper the slope and the greater the elevation gain, the stronger the nighttime breezes. Gravity winds are stronger when the slopes above are bare rock or snow, and weaker when those slopes are covered by vegetation, especially trees. The vegetation slows the overnight cooling that produces gravity winds, and trees deflect and slow the wind. If you are camping at the base of a cliff or in a valley, try to put some trees or other obstacles between your tent and the slope above.

To summarize, gravity winds are strengthened by: 1) clear skies for maximum overnight cooling, 2) prevailing winds of less than 10 miles per hour (16 km/h), and 3) steep slopes of bare rock or snow.

Foehn or Chinook Winds

When winds descend a slope, air temperatures can increase dramatically in what is called a *foehn wind* or, in the western United States, a *chinook* (fig. 36). These winds are significant because of their potential speed, the rapid rise in air temperature associated with them, and the potential they create for rapid melting of snow and flooding.

Chinooks are especially strong over the eastern slopes of the Cascades in Oregon and Washington, as well as over the eastern slopes of the Coast Mountains of British Columbia. The warming is caused by warmer air moving in from the Pacific, rising up the western slopes, crossing the mountain crest, and then sliding down the eastern slopes of the mountains. This air heats as it sinks and compresses on the east side of the crest (fig. 36). Chinooks also destroy the temperature inversion that

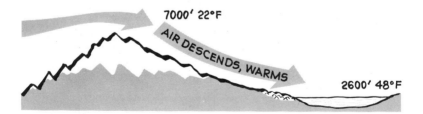

Fig. 36. *Chinook-type winds.*

normally precedes them. Temperatures can easily rise 30 degrees Fahrenheit in minutes, melting as much as a foot of snow in a few hours.

For quick reference, if: 1) you are downwind of a major ridge or crest, primarily to the east of the mountains, and 2) wind speeds across the crest or ridge exceed 30 miles per hour (48 km/h), and 3) precipitation is observed over the crest, then expect a chinook with warming of as much as 6 degrees Fahrenheit per 1,000 feet of descent (3.5°C per 304 m).

For example, the mountains just to the north of Mount Jefferson in central Oregon average 7,000 feet (2,133 m) in elevation. Given an air temperature of 22 degrees Fahrenheit (−5°C), by the time that air reaches aptly named Lake Chinook on the east side of the range at an elevation of roughly 2,600 feet (792 m), the temperature of the air will have warmed to 48 degrees (8.9°C).

The opposite of a chinook is a *bora,* or, as it's called in Alaska, a *taku.* A bora is simply a wind bringing air that is so cold that its sinking, compressing motion as it flows downslope fails to warm it significantly. Such winds are most common downslope of large glaciers. Their speeds can easily exceed 50 miles per hour (80 km/h), especially in Alaska, where the glaciers are huge. Although large glaciers do exist in the mountains of the Pacific Northwest, the boras found in this region don't match those found in Alaska.

The high wind speeds of both chinooks and boras can be at best an inconvenience to hikers, climbers, snowshoers, or skiers, and at worst a danger. Chinooks present additional risks: the rapid warming of snow can increase the risk of avalanches, weaken snow bridges, and lead to sudden flooding from rises in stream levels. When conditions are ripe for a chinook, avoid the downwind slopes.

Light Wind

Certain conditions are conducive to light wind. Expect winds less than 10 knots when you're not at the base of a snowfield, cliff, or mountain and 1) high pressure has moved into the area, 2) pressure changes are less than 1 millibar or .04 inch over a 3-hour period (or your altimeter, when remaining at the same elevation, shows a fluctuation of less than 40 feet [12 m]), and 3) skies are covered by flat, stratus clouds.

One parting bit of advice that may lead to a toastier evening: When winds are light and skies are clear, don't camp at the bottom of a valley. Remember that cold air drains downslope, making campsites upslope actually warmer overnight than those at the base of a mountain or in a valley!

CHAPTER 6

How to Get a Solid, Confidence-Building Weather Briefing Before a Trip

◆

Pre-Trip Weather Briefings

Over the course of the past five chapters, we've covered many guidelines to assist climbers, hikers, snowshoers, and skiers in enjoying the mountains safely. These guidelines are worthless, however, without a foundation of good, current weather information.

As a flight instructor, I teach and expect my students to obtain thorough weather briefings and analyze the effects of the forecasts and current weather information on their routes before setting out. This is no less important for someone heading into the mountains. If the weather deteriorates, a pilot can move away at speeds of 100 miles per hour (160 km/h) or more. The speed of mountain travelers is considerably slower, of course, increasing the hazard of exposure.

There are many good sources of pre-departure weather information. This chapter will review those sources and suggest ways to analyze the information for safe and effective decision making.

Mountain Weather Information Sources

Whether you're planning a picnic or a climb of Mount Rainier, the sources used for gathering weather information and the depth of the research are often the same. Television, radio, and newspapers are the most widely used sources of weather information for the public. Unfortunately, the most widely used sources aren't necessarily the best. Newspaper weather information, although offering you the option of rereading and digesting the data, is dated and occasionally obsolete by the time it reaches your front doorstep or the newsstand. Radio and television weather information is often aggravatingly brief, occasionally with more emphasis placed on cuteness than on content. With some exceptions, recreational weather information tends to be glossed over, even nonexistent. Before relying on a television weather presentation, be certain the weathercaster is a meteorologist with a degree in the field. That's your best guarantee that the information will have been updated to account for changes in the weather. Even when solid information and forecasts are presented, the listener or viewer has to be listening or watching at that exact time. And once the program moves on to other matters, the information is gone.

There are alternatives that every outdoorsperson should be aware of. Tune in to public television's "A.M. Weather" to get a general sense of prevailing weather, likely changes, winds aloft as well as at sea level, and freezing levels. It's a 15-minute presentation that covers the entire United States. Although it's aimed at pilots, the information is developed and delivered by meteorologists, not media models or comedians. "A.M. Weather" offers information applicable to activities in the mountains as well as above them, especially the freezing level, and winds at various elevations above sea level — 3,000 feet, 6,000 feet, and so forth.

The program has two major drawbacks: it airs just once a day, and most public television stations put it on early in the morning. Schedules vary from station to station, so check your local PBS station's program schedule. If you have a VCR, schedule it for automatic recording, then watch the program once you get up. You can do the same for other television weathercasts.

A second good source is NOAA weather radio. NOAA is the National Oceanic and Atmospheric Administration, the parent organization of the National Weather Service. Commonly used by boat owners, good radios dedicated to the frequencies used by NOAA weather radio (no tuning is necessary) can be purchased for less than a hundred dollars and represent good insurance. The best sources for such radios include marine supply and electronics stores.

NOAA weather radio offers regular updates on weather throughout the region, as well as forecasts and warnings of impending severe conditions. Keep one of these radios in your home or your office, or wherever you spend most of your time prior to heading up into the mountains. Given the availability of relatively cheap, compact, hand-held VHF radios, consider carrying one in your pack with extra batteries. Listening usually drains the battery less than transmitting does, so checking NOAA weather radio broadcasts on VHF channels 1 or 2 a couple of times each day shouldn't require a battery change over the course of a two- or three-day trip. The battery may even last as long as a week. Carrying a VHF radio also gives you the option of radioing for assistance if you get in trouble.

A third excellent source of information is the U.S. Forest Service Avalanche Center, which offers news on mountain weather trends as well as avalanche conditions and forecasts during the late autumn, winter, and early spring months. The recorded forecast is available by calling (206) 526-6677 and covers the Olympics, and the Cascades of Oregon and Washington; a similar service is being developed for British Columbia. Calling this number should be considered a must for anyone heading into snow country.

Computerized Weather Data Sources

The computer has revolutionized the workplace and, to some extent, even outdoor recreation. Data banks on trail locations, conditions, and suitability are now available at some outdoor-equipment stores. Computer users also have access to weather data sources ranging from general services such as CompuServe to specific weather information services. The beauty of such sources is that the information is always there; users can get it when they need it or want it, not just when someone else decides to make it available. It's an excellent way to get a thorough picture of present and future weather before a trip.

Expect the area of computer weather data to change rapidly with expanding services. Such change makes it impractical to attempt to offer a definitive listing of computer weather data sources. Appendix 5 lists a few worth checking into.

There's another source, which should be used with restraint but which can be very helpful in a jam. The Flight Service Station briefers who prepare pilots are willing, on a time-available basis, to assist mountaineers. There are four rules that should be followed to prevent such willingness to extend assistance from disappearing:

- Call during slow times: from 8 P.M. to 5 A.M. or 10 A.M. to 2 P.M.
- Give the briefer a specific list of needs. Don't waste his or her time with vague questions such as, "What's it look like for a climb of Mount Hood tomorrow?" This should be a last-resort source of information on winds, freezing levels, etc.
- Remember, the briefer can give information only, not advice. It's up to you to make a decision.
- For purposes of the briefer's record-keeping, be prepared to provide a listing of four numbers followed by a letter, a format that matches an aircraft identification (example: 1234 Alpha, using the phonetic alphabet). The briefer records this listing so that time spent on briefings will be credited in budgeting money and staff.

To contact an FAA Flight Service Station briefer, call 1-800-WX-BRIEF. Remember, abusing this service will result in the loss of a valuable option not only to you, but also to your fellow mountaineers.

Home-Based Weather Observations

A barometer is one tool you shouldn't be without in your home. It offers valuable information and guidance on likely winds in the mountains and the speed at which a weather disturbance is approaching or at which a high is building in. Choose one with a resettable guide, which will allow you to quickly see how much the pressure has risen or fallen since you last checked. Here are some guidelines for measuring over a 3-hour period. Pressures are given in inches of mercury or in millibars.

Pressure Fall Guides for Three-Hour Period

- *If* less than .04 inch/1.2 mb, *then* monitor as usual.
- *If* .04–.06 inch/1.2–1.8 mb, *then* watch sky and winds closely.
- *If* .07–.08 inch/1.8–2.4 mb, *then* monitor weather radio continuously and consider delaying backcountry trip.
- *If* more than .08 inch/2.4 mb, *then* delay departure.

Incidentally, a pressure rise is no guarantee of wonderful weather here in the Pacific Northwest. If you get reports of a high within 600 to 700 miles (960–1,120 km) of the coast with a low in eastern Washington, eastern Oregon, southeastern British Columbia, or northern Idaho, and you note a 3-hour pressure rise of 2 to 4 millibars, then gale-force winds are possible, that is, wind speeds exceeding 33 knots. Given the channeling effect of passes and gaps in the terrain, speeds of 60 knots in the mountains can develop. Consider waiting until the high has had an opportunity to build into the Northwest; that will usually happen within 12 to 24 hours.

Also, keep an eye on the sky for changes in cloud cover, specifically 1) lenticulars over the peaks — a sign of high winds at altitude and possibly an approaching low, and 2) a halo around the sun or moon — a sign of an approaching disturbance within as few as 12 and possibly as many as 36 hours, if followed by thickening, lowering clouds.

Home weather stations can be fun and informative. Some even connect to personal computers to store weather data. Popular models are manufactured by Radio Shack and Rain Wise. Look for them at outdoor-equipment stores and nautical supply firms.

At the very least, a good weather vane on top of your roof can offer some good guidance, if it's not obstructed by high buildings or trees. Remember these points when monitoring your weather vane:

- A shift from north or northeasterly winds to southeasterlies generally signals the approach of a low.
- A shift from southeast to southwest usually signals the passage of a cold front.
- A shift from southwest to north usually indicates the building of high pressure.

Information Gathering with a Plan

Given all these sources of information, the key is to use them well. Information by itself is of limited use; information gathered with a purpose is of great use.

To best assess the suitability of weather in the mountains, follow a plan. The following checklist is one such plan. Use it, modify it, develop one better suited to your needs. The key is to organize your weather information gathering.

Mountain Weather Checklist

Planned Trip: _____ **Date(s):** _____

Checklist

___Synopsis
___Warnings/Watches
___Winds Aloft
___Freezing/Snow Level
___Zone Forecast(s)
___Extended Forecast
___Avalanche Forecast
___Field Altimeter/Barometer
___Compass
___Notebook/Logbook

Preferred Sources

PBS A.M. Weather/Channel
NOAA Weather Radio
  VHF Channels 1 & 2
Avalanche Hotline
  (206) 526-6677
Computer Data Bank
  Tel.# _____
  User ID _____
  Password _____

Weather Synopsis: _____

_____

_____

_____

_____

_____

Warnings/Watches: _____

_____

_____

_____

_____

_____

Winds Aloft:	Alt.	Wind Dir.	Wind Speed	
	3,000 ft.	_____		Freezing/Snow Level(s)
	6,000 ft.	_____		
	9,000 ft.	_____		_____
	12,000 ft.	_____		_____
	15,000 ft.	_____		_____

Zone Forecast(s)

	Date: _____	Date: _____	Date: _____
Sky	_____	_____	_____
Cloud Bases	_____	_____	_____
Cloud Tops	_____	_____	_____
Temps	_____	_____	_____

Extended Outlook from _____ to _____

Avalanche Outlook: _____

_____

_____

_____

Additional Notes: _____

_____

_____

_____

_____

Observations: _____

_____

_____

_____

_____

_____

_____

_____

_____

_____

Planning Map:
   (Sketch or photocopy of your route map)

Using a Plan Effectively

The best way to use the plan outline is not to wait until just a few hours before you leave. Obviously, some information is better than no information, but consider getting weather data at least one day and preferably two days before your planned departure. That will give you a chance to verify the forecasts with observed conditions. If the forecasts are pretty close to what's observed, planning can proceed with more confidence than if the forecast and observed weather conditions are 180 degrees apart. Here's a suggested sequence for gathering information:

Two Days Before the Trip
   1. Large-scale pattern
   2. Projected weather for the next two days

One Day Before the Trip
   1. Current weather to evaluate previous day's forecasts
   2. Large-scale weather pattern
   3. Projected weather for the next two days

Day of the Trip
   1. Current weather to evaluate previous day's forecasts
   2. Projected weather for the trip

Let's take an example to see how such a plan can be used effectively:

Mountain Weather Checklist (Sample 1)

Planned Trip: Camp Muir on Mount Rainier   **Date(s):** 7/11 up
            From 5,000 to 10,000 feet   7/12 down

Checklist

_x_ Synopsis
_x_ Warnings/Watches
_x_ Winds Aloft
_x_ Freezing/Snow Level
_x_ Zone Forecast(s)
_x_ Extended Forecast
_x_ Avalanche Forecast
_x_ Field Altimeter/Barometer
_x_ Compass
_x_ Notebook/Logbook

Preferred Sources

PBS A.M. Weather/Channel 9
NOAA Weather Radio
   VHF Channels 1 & 2
Avalanche Hotline
   (206) 526-6677
Computer Data Bank
   Tel.#: 747-1000
User ID: 12345
Password: HIRNHIR

Weather Synopsis: Cold front due to move through late evening hours of 7/10, then high pressure expected to build along the coast.
(*Source:* NOAA weather radio)

Warnings/Watches: Gale warnings posted along the coast, Strait of Juan de Fuca, Small Craft Advisory in Puget Sound.
(*Source:* NOAA weather radio)

Winds Aloft:

Alt.	Wind Dir.	Wind Speed	
3,000 ft.	160	30 knots	Freezing Level: 10,000 ft.
6,000 ft.	170	35	forecast to drop to
9,000 ft.	200	38	7,000 ft. 7/10
12,000 ft.	220	44	
15,000 ft.	230	50	

(*Source:* FAA Flight Service)

Zone Forecast(s): Seattle

	Date: 7/9—A.M.	Date: 7/9—P.M.	Date: 7/10—A.M.
Sky:	Ptly to mstly cldy	Cldy/rain	Rain—showers
Cloud Bases:	8,000 ft.	N/A	N/A
Cloud Tops:	15,000 ft.	N/A	N/A
Temps:	70–77	60–63	57–63

(*Source:* NOAA weather radio/FAA Flight Service for cloud bases/tops)

Extended Outlook from 7/11 to 7/13: Showers ending 7/11 with partial afternoon clearing. Mostly sunny after low morning clouds 7/12. Mostly sunny 7/13.

(*Source:* NOAA weather radio)

Avalanche Outlook: Not available

Additional Notes: Barometer at home showing slow, but steady downward trend.

Observations: Forecasts have been a little off. One system after another has zinged through, but we've enjoyed a little clearing. The forecast for later today (7/9) seems a little pessimistic. The clouds have been slower to move in than expected. As of this A.M., surface winds are still out of the NE, with only high cirrus moving in. There was a wide halo around the moon last night (7/8), and a mousy-looking cap cloud over the summit of the mountain.

Planning Map: (Photocopy of route map)

Analysis: The pattern described is typical during the summer in the Northwest: the hint of sunshine interspersed with clouds, precipitation, and wind. The forecasters are having their difficulties with the arrival times of the systems, but the overall trend seems good. Short-term, the forecast calls for an approaching low to bring more clouds and rain, with the freezing level dropping to approximately 7,000 feet the day before the planned overnight hike to Camp Muir at 10,000 feet.

Even though the long-range outlook calls for high pressure and clearing, the early arrival of a cold front should set off some mental alarms on two counts: first, the later-than-forecast approach of the next weather system, and second, the forecast arrival of cold air aloft, which will drive down the freezing level.

The later-than-forecast approach of the next weather system probably shouldn't be viewed as a total reprieve; this is more a stay of execution. Although the forecast issued the day this checklist was written was overly pessimistic, there are several indicators pointing toward the eventual arrival of the system.

First, the existence of Gale Warnings along the coast and strait and Small Craft Advisories in Puget Sound bears out the logged observation of a sustained pressure drop in suggesting the approach of a low.

Next, the cap cloud forming over the summit of Mount Rainier indicates the approach of moisture in the upper atmosphere, as does the appearance of a ring around the moon the evening before. The fact that it was a wide ring suggests that precipitation was at least 24 and possibly as much as 48 hours distant at the time of the observation (the evening of July 8), placing the likely time of cold-front passage late on July 10, instead of overnight on the 9th, or in the early morning hours of the 10th.

As a result, the planned hike to Camp Muir early on July 11 could very well run into lingering snow showers from 6,000 feet on up. (Remember, the snow level is usually 1,000 feet [304 m] below the freezing level.) Another consideration is that, given the push of colder air into the region from across the Pacific, unstable air will be running up against Mount Rainier, where it will continue to rise, setting off orographic rain showers well after the cold front passes. Therefore, it seems the planned hike to Camp Muir in the example is a little iffy at best. A good alternate would be to select a hike east of the Cascades.

Mountain Weather Checklist (Sample 2)

Planned Trip: Hike on Mount Si 600–4,200 feet   **Date(s):** 9/15

Checklist	Preferred Sources
_x_ Synopsis	PBS A.M. Weather/Channel 9
_x_ Warnings/Watches	NOAA Weather Radio
_x_ Winds Aloft	VHF Channels 1 & 2
_x_ Freezing/Snow Level	Avalanche Hotline
_x_ Zone Forecast(s)	(206) 526-6677
_x_ Extended Forecast	Computer Data Bank
_x_ Avalanche Forecast	Tel.#: 747-1000
_x_ Field Altimeter/Barometer	User ID: 12345
_x_ Compass	Password: HIRNHIR
_x_ Notebook/Logbook	

Weather Synopsis: A surface low is expected to pass through southern British Columbia in the early afternoon hours of 9/14, the day before the planned hike. The cold front trailing from that low is forecast to pass through Puget Sound during the mid- to late-afternoon hours of 9/14.

High pressure will gradually build into Puget Sound overnight, with clearing forecast 9/15.
(*Source:* NOAA weather radio)

Warnings/Watches: Small Craft Advisory issued for Strait of Juan de Fuca, Admiralty Inlet, Puget Sound, Hood Canal, and from Camano Island north to Point Roberts.
(*Source:* NOAA weather radio)

Winds Aloft:

Alt.	Wind Dir.	Wind Speed	
3,000 ft.	130	10 knots	Freezing/Snow Level(s)
6,000 ft.	145	15	
9,000 ft.	170	25	Freezing Level: 7,000 ft.
12,000 ft.	190	40	
15,000 ft.	230	50	

(*Source:* FAA Flight Service)

Zone Forecast(s): Seattle

	Date: 9/14	Date: 9/15	Date: 9/16
Sky:	Cloudy/rain	Partly cloudy	Mostly sunny
Cloud Bases:	2,000 ft.	4,000 ft.	N/A
Cloud Tops:	N/A	N/A	N/A
Temps:	55–60	52–57	60–65
Winds:	S 5–20 kn	S-SW 5–15 kn	N 5–15 kn

(*Source:* NOAA weather radio/FAA Flight Service)

Extended Outlook from 9/17 to 9/19: Mostly sunny with patchy morning fog, overnight lows in the 40s, daytime highs in the 60s.
(*Source:* NOAA weather radio)

Avalanche Outlook: Not available/not applicable

Additional Notes: Clouds have been thickening and lowering, local winds picking up slightly out of southeast. Barometer dropping steadily. Looks like rain will start anytime.

Observations: Forecast timing has been right on the last couple days. Last-minute check of the recorded Flight Service Station observations on the morning of 9/15 did show partial clearing and no precipitation, but winds in south Puget Sound at Olympia and Tacoma were out of the south to southwest, while those in Seattle were out of the west, and those at Everett were out of the northwest. Strange.

Analysis: Generally, the forecast issued gives good reason for optimism, especially since the preceding forecasts have been correct. General observations the next morning, the day of the planned hike on Mount Si, seem to support the forecast of partly cloudy skies and dry weather. However, the quirky winds noted in the Puget Sound area that morning should raise some caution flags. The wind pattern reported is very suggestive of the Puget Sound Convergence Zone.

Remember, the convergence zone typically follows cold-front passage, with high pressure building in along the coast. A southwest-to-northwest flow along the coast splits around the Olympics, converging in central Puget Sound, usually between Everett and Tacoma. Flight Service Station or NOAA weather radio forecasts reporting south to southwest winds at Tacoma and Olympia indicate air flowing northward after moving around the southern end of the Olympics, while northwest winds reported at Everett indicate air flowing southward after moving around the northern end of the Olympics. The westerly winds at Seattle probably are caused by those two opposing currents of air colliding.

The collision of air from the north and south that gives the convergence zone its name results in rising air, and often cloud formation, or thicker clouds than might be found elsewhere. Also, showers typically occur, and occasionally thundershowers, even when not anticipated in the forecast.

Since weather conditions strongly point to the formation of the Puget Sound Convergence Zone and the possibility of showers or even thundershowers, it would be wise to choose an alternate, since Mount Si is right in the midst of the convergence zone. Select a site to the north of Everett, north of Mount Pilchuck, or to the south of Tacoma, south of Mount Rainier.

Mountain Weather Checklist (Sample 3)

Planned Trip: Climbing at Smith Rock, Oregon   **Date(s):** 4/22

Checklist

- _x_ Synopsis
- _x_ Warnings/Watches
- _x_ Winds Aloft
- _x_ Freezing/Snow Level
- _x_ Zone Forecast(s)
- _x_ Extended Forecast
- _x_ Avalanche Forecast
- _x_ Field Altimeter/Barometer
- _x_ Compass
- _x_ Notebook/Logbook

Preferred Sources

PBS A.M. Weather/Channel 9
NOAA Weather Radio
   VHF Channels 1 & 2
Avalanche Hotline
   (206) 526-6677
Computer Data Bank
   Tel.#: 747-1000
   User ID: 12345
   Password: HIRNHIR

Weather Synopsis: As of the evening of the 21st, high pressure currently dominates the region but is gradually receding to the southeast. A 985-millibar low is approaching the Oregon Coast, with an occluded front expected to cross the Cascades tomorrow (4/22).
(*Source:* NOAA weather radio)

Warnings/Watches: Gale Warnings are posted for the Oregon coast.
(*Source:* NOAA weather radio)

Winds Aloft:

Alt.	Wind Dir.	Wind Speed	
3,000 ft.	120	10 knots	Freezing/ Snow Level(s)
6,000 ft.	145	25	9,000-foot freezing level
9,000 ft.	195	30	
12,000 ft.	215	55	
15,000 ft.	240	80	

(*Source:* FAA Flight Service)

Zone Forecast(s): Seattle

	Date: 4/21	Date: 4/22	Date: 4/23
Sky:	Increasing clouds	Cloudy, rain late evening	Rain, changing to showers
Cloud Bases:	22,000 ft.	N/A	N/A
Cloud Tops:	N/A	N/A	N/A
Temps:	47–52	52–57	45–50

(*Source:* NOAA weather radio/FAA Flight Service)

Extended Outlook from 4/24 to 4/26: Mostly sunny, breezy

Lows: 30–37

Highs: 50s

(*Source:* NOAA weather radio)

Avalanche Outlook: N/A

Additional Notes: The timing of the approaching low has changed from forecast to forecast. On the evening of the 20th, forecasters expected the low to cross the Cascades early on the 22nd. The next morning, on the 21st, the timing was changed to bring the front through much later, on the morning of the 23rd. Now the timing has been changed again, with frontal passage expected on the evening of the 22nd. Also, your barometer has been falling rapidly, showing a pressure drop of 1.5 inches since morning (8 hours ago). A tight corona encircles the moon from cirrus clouds.

Analysis: Anytime the forecast timing fluctuates, be suspicious. In this case, the low was probably approaching from the west to southwest, and forecasters expected (probably with good reason) that it would continue approaching Oregon at the same speed. Suddenly it slowed, probably intensifying, leading to a change in the forecast time of frontal passage. A low with a central pressure of 985 millibars is a fairly intense low. With the intensification process completed, the low has speeded up, leading to yet another revision of the forecast. Don't placidly accept that the front will approach the Smith Rock area the evening of the 22nd.

It might seem that the forecast time of frontal passage during the evening hours of the 22nd would allow plenty of time for climbing on Smith Rock. However, the drop in air pressure shown on the home barometer is fairly steep, and the winds aloft are strong, 80 knots at 15,000 feet. In addition, the fact that a corona has been observed encircling the moon suggests the possibility of an earlier-than-expected frontal passage. Typically, a corona indicates that precipitation is likely to begin within 12 to 24 hours. As the climber is checking weather conditions on the evening before, it's very possible the precipitation could begin during the mid- to late-morning hours of the 22nd, with frontal passage possible during the early- to mid-afternoon hours, right during the climbing.

My suggestion would be to get up a little earlier than planned and check the current weather conditions and the latest forecast. If wind speeds are picking up and cloud bases are lower than 10,000 feet (3,048 m), expect the rain to begin within a couple of hours, and a very forceful frontal passage not long after that. Thunderstorms are possible given the drop in temperature and the strong winds aloft. If wind speeds haven't picked up and cloud bases are above 12,000 to 15,000 feet (3,657–4,572 m), you may have time to get a quick climb in. During the climb, keep a sharp eye on the clouds, wind speeds, and, if you can, on your pocket altimeter/barometer.

◆

The preceding examples show how personal observations, paired with official weather observations and forecasts and a measure of weather know-how, can produce a good decision. This doesn't preclude the possibility of a surprise in the field; that's the reason for the next and final chapter, a summary of field forecasting hints for the mountains of the Pacific Northwest.

CHAPTER 7

Field Forecasting Guidelines

◆

A climbing party slowly begins its ascent of Mount Baker in the pale predawn light, muscles burning, eyes still blurry. The thin wisps of icy cirrus clouds admired by the group of young people and their guides the evening before have broadened into a leaden shield of altostratus. Faced with conflicting weather forecasts, the group's optimists prevail with the argument that rescheduling the climb would be difficult, if not impossible.

One of the guides notices that her pocket altimeter is reading higher than the actual base-camp altitude. But with inexperienced climbers demanding help in strapping on crampons, she resets the altimeter to the proper reading without further analysis.

Three hours later, thick, gray clouds veil the summit from view. Wet snow begins to fall, crusting over mustaches, beards, even eyelashes. As the group moves slowly up the mountain, the cold air left over from the early morning is replaced by driving southeasterly winds. The force of the gale bends the wands marking the climbing route. What began as snow changes first to a mixture of rain and snow, and then to rain. The snow underfoot takes on the consistency of oatmeal, a thick slush that clings to crampons. The progress of the group slows to a crawl, and the guides begin to consider ending the climb. Above and to the right, a low rumble grows steadily louder. Visibility is too poor to confirm the source, but the experienced ears of the guides tell them it's an avalanche. The group heads back to base camp.

The scenario above combines elements of several climbing case histories. Weather-related climbing accidents rarely occur without any warning. Occasionally the clues are subtle, at other times broad as daylight. Caution can't end with the decision to go; anyone who has spent time in the backcountry knows that weather forecasts can and do go wrong.

Even the best forecasters have their "off days"; many of the busted forecasts aren't due to human error, but are the result of limited computer ability to digest a mountain of weather observations, run those observations through complicated equations, and develop projections. The computer forecast models now in use are immeasurably better than those used just a decade ago, but they are gluttons for computer memory. Whenever a new supercomputer is developed, its capability is often tested by running weather forecasting models.

As impressive as these computers are, they can't fit in all the data necessary to accurately assess the evolution of a weather pattern in the

Pacific Northwest. As of this writing, the coastline wasn't accurately located, and the Olympics and Cascades weren't factored into the models (the computer only pictured a gradual upslope from the West Coast to the Rockies). Connect an airplane's autopilot to the computer model, and it would very confidently fly due east from the coast at 3,000 feet (914 m), directly into the Olympics.

Previous chapters have shown the wide variations in weather over short distances in the Pacific Northwest caused by variations in terrain. Another problem faced by the computer models is that they are a little like fishnets, with observing stations forming the intersections between strands, the mesh. Just as fish smaller than the mesh can swim through the net, so can weather disturbances smaller than the "mesh" of the computer weather models also slip through unnoticed.

Forecasters experienced in a region's weather patterns can on occasion correct errors in the computer models' forecasts, but not always. That's why it's important for hikers, climbers, skiers, and snowshoers — in fact, anyone who plays or works more than easy walking distance from safe shelter — to understand weather and to be able to do some field forecasting on their own. Pointing a finger at the forecaster or the computer models won't save your bacon on the side of a mountain!

In the story at the beginning of this chapter, for example, the climbing guides had several early clues that pointed toward deteriorating weather. Thickening and gradually lowering clouds and an altimeter that suddenly showed a higher altitude (indicating rapidly falling pressure) both indicated an approaching storm. The gradual change from dry snow to wet snow suggested an approaching warm front, and the sudden increase in wind speed should have convinced even the most optimistic of climbers.

This chapter won't break new ground so much as it will reorganize the principles previously discussed into flowcharts to guide your decision making. Some of the guidelines from previous chapters will be repeated here. This is the section you'll want to refer to in the backcountry. It is no substitute for current forecasts and observations; like ski bindings or climbing gear, field forecasting guidelines don't replace good judgment, they enhance it. The flowcharts in this chapter move from guidelines associated with large-scale weather systems to more narrowly defined situations relating to specific seasonal or terrain-influenced patterns. Check the first box, and if it applies, continue to the next one in the flowchart series.

— How to Watch for Major Weather Systems —

There are four major indicators of an approaching storm: changes in cloud cover, in wind, and in air pressure. No single indicator is infallible; all three should be examined.

Cloud Cover Clues

IF	THEN	
High cirrus clouds forming loose halo around sun/moon	Precipitation possible within 24–48 hours	Watch for lowering, thickening clouds
High cirrus clouds forming tight corona around sun/moon	Precipitation likely within 24 hours	Watch for lowering, thickening clouds
"Cap" or lenticular clouds forming over peaks	Precipitation possible within 24–48 hours	Watch for lowering, thickening clouds
Thickening, lowering layered/flat clouds	Warm/occluded front likely approaching 12–24 hours	Check for wind shifts, pressure drops
Breaks in cloud cover closing up	Cold front likely within 12 hours	Check for wind shifts, pressure drops

Clues from Winds Aloft

IF	THEN		
Cirrus clouds moving from SSW to NNE	Expect almost continuous moderate to heavy precipi-	tation	3 in. to 4 in. precipitation on western slopes
Cirrus clouds moving from WSW	Expect moderate precipitation with breaks between systems	Thunderstorms possible along western slopes	
Cirrus clouds moving from WNW	Fast-moving cold front with moderate precipitation	Thunderstorms likely within 48 hours of cold-front passage along western slopes	
Cirrus clouds moving from north, Dec.– Feb.	Freezing level at/near sea level	Snow likely	Gusty, cold winds out of Fraser River first, then passes further south. Velocities 60–100 kn possible

IF	THEN	
Cirrus clouds moving from NNE	Sunny weather likely, with above-normal temperatures in summer, below-normal in winter	Temperature inversion possible in two or more days, with lowered visibility in basins

Clues from Surface Winds

Refer to figures 21, 22, and 23, a series of maps showing peak precipitation areas associated with different surface-wind directions.

Clues from Pressure Changes

Remember that a pocket altimeter can give excellent indications of an approaching weather system. An altimeter that registers an increase in altitude, even though none has taken place, is actually reporting a drop in air pressure. For that reason, the following chart correlates changes in pressure and in altitude readings. As with all guidelines, the summary is no substitute for obtaining a complete forecast, but it can provide a useful warning aid.

Changes in pressure create changes in wind and are often related to approaching fronts that may bring precipitation. Notice that pressure indications are given both in inches of mercury and in millibars, while the altimeter readings are given both in feet and in meters.

Air Pressure Drop or over 3 Hours	Altimeter Rise over 3 Hours	Advised Action
Less than .02 to .04 in. Less than 0.6 to 1.2 mb	Less than 20 ft. Less than 6 m	None — continue to monitor
.04 to .06 in. 1.2 to 1.8 mb	40 to 60 ft. 12 to 18 m	Cloud lowering, thickening? Check hourly
.07 to .08 in. 1.9 to 2.4 mb	70 to 80 ft. 19 to 24 m	Consider ending outing, find safe bivouac site
More than .08 in. More than 2.4 mb	More than 80 ft. More than 24 m	Find safe bivouac site now!

Other Weather Indicators

Fog Indications

Fog is rarely, if ever, life-threatening. But it can make route-finding difficult and enjoyment of scenery impossible. With a little knowledge and careful observation, fog is relatively easy to anticipate.

IF	AND IF	THEN
Ground wet, pressure increasing after frontal passage?	Sky clearing overnight, winds 5 kn or less	*Radiation fog* likely by or shortly after sunrise. Expect burnoff by noon unless inversion present.
Location Olympics, Coast Range, western slopes Cascades/ Coast Mtns.	Above-average temperatures over region during late spring, summer; cooler air, gusty westerly winds in afternoon/evening, followed by stratus clouds	*Advection fog* next morning on western slopes, possible drizzle
Location east of Cascades/ Coast Mtns.	Increasing pressure, cooler air moving in	*Upslope fog* likely on windward slopes, burnoff possible in afternoon, evening hours.
Thickening, lowering clouds, falling pressure	Rain falling into cold air, especially in valleys	*Warm frontal fog* likely. Won't clear until after warm-front passage.

Snow Indicators

Snow vs. rain vs. freezing rain — which of these occurs can make an immense difference in the outcome of a trip. Following are some guidelines to assist in making projections in the field.

Knowing the actual or forecast freezing level is the basis for such guesstimates. Without that knowledge, it's possible to estimate the freezing level by using the standard lapse rate. However, the standard lapse rate rarely exists in the atmosphere, so allow for a margin of error of 500 to 1,000 feet (152–304 m). Here's a refresher on the calculations explained in Chapter 4.

Estimating the Freezing Level
(in feet and degrees Fahrenheit)

$$\text{YOUR ELEVATION} + \frac{(\text{CAMP TEMPERATURE} - 32°) \times 1000 \text{ FT.}}{3.5°} = \text{ESTIMATED FREEZING LEVEL IN FEET}$$

Estimating the Freezing Level
(in meters and degrees Celsius)

$$\text{YOUR ELEVATION} + \frac{\text{CAMP TEMPERATURE} - 0° \times 304}{2°} = \text{ESTIMATED FREEZING LEVEL IN METERS}$$

In both cases, you're subtracting the freezing temperature (32 degrees on the Fahrenheit scale, and 0 on the Celsius scale). The numbers used for the division both indicate the standard lapse rate, but from the two temperature scales.

Having obtained or estimated the freezing level, use the following guidelines to estimate the snow level.

IF	AND IF	THEN
Stratus clouds or fog present	Steady, widespread precipitation	Expect to find the snow level 1,000 ft. (304 m) BELOW the freezing level.
Cumulus clouds present or cold front approaching	There is locally heavy precipitation that varies from time to time or place to place	Expect to find the snow level up to 2,000 ft. (609 m) BELOW the freezing level. Snow will stick 1,000 ft. below the freezing level.

Guidelines for Finding Quality Snow

IF	AND IF	THEN	OR
Late October to March	Snow actively falling	Choose elevations with air temperatures of 20 to 25 degrees (–3° to –7°C)	1,000 to 2,000 ft. (304 to 609 m) above the freezing level

Here are a few additional suggestions for maximum enjoyment:

Winds light: Go above timberline
Winds strong: Stay below timberline/on leeside slopes
Morning hours: Travel on south-facing slopes
Afternoon hours: Travel on north-facing slopes
And one final hint: The best snow for recreation comes with a west-northwesterly storm track. If that's in the forecast, cancel all plans and head for the mountains!

Safety Guidelines

Avalanche Guidelines

Avalanche forecasting is a precarious business; witness the number of professional avalanche patrollers and forecasters who have been killed or injured in the line of duty. As discussed in Chapter 4, anyone planning to travel in the backcountry at or above the snow line should take a winter mountain survival course or similar program. The National Ski Patrol offers an excellent one. What follows are simply brief field guidelines to augment training and a thorough pre-trip briefing.

IF	AND IF	THEN	
Coastal Washington, Oregon, or B.C.	Major precipitation within the last 24 hours	Significant avalanche hazard	
Precipitation occurring	Temperatures rising, freezing level rising	Avalanche risk increasing	
Wind speeds 10–15 mph (16–24 km/h) or greater		Avalance risk increasing	
Snow falling at 1 in. (2.5 cm) per hour + for 8 hours or more	Wind speeds 10 mph (16 km/h) +	High avalanche hazard	
Spring/summer months	Daytime temperatures above feezing	Thickening, lowering clouds overnight	Avalanche risk increasing, especially late morning and early afternoon

Thunderstorm and Lightning Safety Guidelines

The best safety advice for thunderstorms is simple: Avoid them. Here are some guidelines to suggest what to do if caught near one.

Start timing at lightning ➡	Stop timing at thunderclap ➡	Divide by 5 for distance from storm in miles

IF	THEN
Interval increasing	Storm moving away
Interval decreasing	Storm approaching

DO watch for cumulus showing strong upward growth.
DO choose a campsite uphill from valley floor.
DO get away from exposed areas, pinnacles, peaks.
DO get inside car or building if possible.
DO get away from water.
DO seek low ground in open valleys, meadows.
DO move at once if hair or scalp feels tingly.
DO NOT stand under trees.

Localized Wind Phenomena

Chinook Winds

IF	AND IF	THEN
Location leeside of major crest	Winds at crest elevation 30 mph (48 km/h) or greater and precipitation over mountains	Expect warming of 6 degrees/1,000 ft. of descent (11.5°C/ 1,000 m)

Gravity Winds

IF	AND IF	AND IF	THEN
Sky clear	Prevailing winds less than 11 mph (17 km/h)	Location along steep slopes with no vegetation	Expect gravity winds moving downslope

Light Winds (Less than 10 knots)

IF	AND IF	AND IF	AND IF	THEN
High pressure dominates	Pressure change less than .04 in. or 1 mb in 3 hours	Sky clear or stratus clouds	Location not at base of snowfield, cliff, or mountain	Expect winds less than 11 mph (17 km/h)

Strong Winds (Greater than 30 Knots)

IF	BETWEEN	THEN
Pressure difference 4 mb or more	Quillayute and Bellingham	Expect strong winds in southern B.C., Washington passes
	Astoria and The Dalles	Expect strong winds in southern Washington, northern Oregon passes
	Astoria and Quillayute	Expect strong winds in interior valleys, coastal region

Gap Winds/Wind Channeling

IF	THEN	
To leeward of pass or gap	Wind will tend to follow land contours	Wind speed may double in pass/gap, near downwind opening

Terrain Blocking

IF	THEN
To leeward of major peak	Prevailing wind will reestablish at a distance of five times the elevation of the peak above the surrounding terrain.

Puget Sound Convergence Zone

IF	AND IF	THEN	
Coastal winds from WSW to WNW	Northerly winds in Admiralty Inlet or southerly in South Puget Sound	Expect Puget Sound Convergence Zone	Thunder, rain, snow showers from Glacier Peak to Mount Rainier. Better weather to north and south of zone.

The preceding guidelines, it must be emphasized, are not intended to replace forecasts; they're meant to augment such forecasts in the field. There is no substitute for an updated forecast from a professional meteorologist; obtaining one should precede every trip into the mountains.

This is a spectacular region in which to enjoy the natural beauty and serenity of mountains. Whether it's excitement, challenge, or relaxation you're after, there are special places for every taste. There's little doubt that at times the weather in the Pacific Northwest mountains adds an element of risk and even danger. But, with the help of knowledge and good judgment, each visitor will have the opportunity to return to those special high places time and time again.

Appendixes

Appendix 1

Wind-Chill Chart
(Wind in Miles Per Hour)

	0	5	10	15	20	25	30	35	40	45	50
	35	33	21	16	12	7	5	3	1	1	0
	30	27	16	11	3	0	-2	-4	-4	-6	-7
	25	21	9	1	-4	-7	-11	-13	-15	-17	-17
	20	16	2	-6	-9	-15	-18	-20	-22	-24	-24
	15	12	-2	-11	-17	-22	-26	-27	-29	-31	-31
	10	7	-9	-18	-24	-29	-33	-35	-36	-38	-38
	5	1	-15	-25	-32	-37	-41	-43	-45	-46	-47
	0	-6	-22	-33	-40	-45	-49	-52	-54	-54	-56
	-5	-11	-27	-40	-46	-52	-56	-60	-62	-63	-63
	-10	-15	-31	-45	-52	-58	-63	-67	-69	-70	-70
	-15	-20	-38	-51	-60	-67	-70	-72	-76	-78	-79
	-20	-26	-45	-60	-68	-75	-78	-83	-87	-87	-88
	-25	-31	-52	-65	-76	-83	-87	-90	-94	-94	-96
	-30	-35	-58	-70	-81	-89	-94	-98	-101	-101	-103
	-35	-41	-64	-78	-88	-96	-101	-105	-107	-108	-110

CURRENT TEMPERATURE

Appendix 2

Cloud Identification Chart

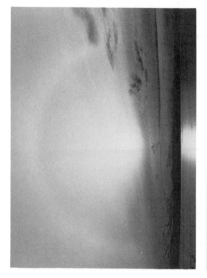

Halo ▶
Commonly seen
24–48 hours ahead
of precipitation

◀ *Cirrocumulus*
Often changes into
cirrus

Lenticular ▶
Wavelike clouds
over mountains
often suggesting
approaching
precipitation with-
in 48 hours

▼ *Stratus*
Layerlike clouds approaching warm front, or ocean air

Cirrostratus ▲
Often indicates approaching warm front

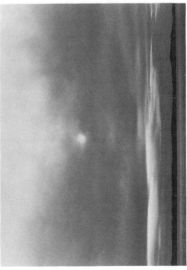

▼ *Altostratus*
When part of approaching warm front, follows cirrostratus

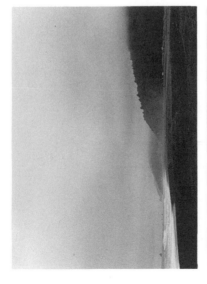

Nimbostratus ▲
Stratus clouds producing widespread precipitation and low visibility

▼ Cumulus
With continued upward growth these suggest showers later in the day

Altocumulus ►
High-based clouds often indicating potential for thunder, rain showers

◄ Cumulonimbus
Cumulus producing rain, snow, or thunder and lightning

Stratocumulus ►
Lumpy, layered clouds often following a cold front, suggesting showers.

Appendix 3

Suggested Reading

Gedzelman, Stanley D., *The Science and Wonders of the Atmosphere*. New York: John Wiley & Sons, 1980.

Graydon, Don, ed., *Mountaineering, The Freedom of the Hills*. 5th ed. Seattle: The Mountaineers, 1992.

LaChapelle, Edward R., *The ABC of Avalanche Safety*. Seattle: The Mountaineers, 1978.

LaChapelle, Edward R., *Field Guide to Snow Crystals*. Seattle: University of Washington Press, 1969.

U.S. Department of Agriculture. Forest Service. *Avalanche Handbook*. Agricultural Handbook no. 489, 1976.

U.S. Department of Transportation. Federal Aviation Administration. *Aviation Weather*. Advisory Circular no. 00-6A, 1987.

U.S. Department of Transportation. Federal Aviation Administration. *Aviation Weather Services*. Advisory Circular no. 00-45C, 1987.

Wilkerson, James A., M.D., ed., *Hypothermia, Frostbite and Other Cold Injuries*. Seattle: The Mountaineers, 1982.

Appendix 4

Sources of Weather Equipment and Information

Radios
Fisheries Supply, Seattle, Wash. (800) 562-6999 (in Washington); (800) 426-6930 (elsewhere).

Weather Instrumentation/Equipment
Alden Electronics, Westborough, Mass. (408) 366-8851.
Fisheries Supply (see above)
Sensor Instruments Company, Inc., Concord, Mass. (800) 633-1033.
Wind & Weather, Mendocino, Calif. (800) 922-9463.

Weather Data
Accu-Weather, State College, Pa. (814) 234-9601.
Quorum Communications, Grapevine, Tex. (817) 488-4861.
WeatherFax, Maynard, Mass. (800) 359-4242.
Weather Information Systems, Santa Paula, Calif. (805) 933-1270.
WSI Corporation, Bedford, Mass. (617) 275-5300.

Appendix 5

Computer Weather Data Sources

Here is a partial listing of available weather data sources. Some provide not only scripted weather forecasts and data, but also weather charts. A few are heavily weighted toward the aviation user, but it's these very services that offer the information usually missing elsewhere, namely, winds above sea level and the freezing/snow level. To help you understand the forecast and observation abbreviations and formats, pick up a copy of FAA Advisory Circular 00-45C at any airport facility offering flight training for pilots, or at a Government Printing Office bookstore.

COMPUSERVE
PO Box 20212
5000 Arlington Center Blvd.
Columbus, OH 43220
(800) 848-8199
(614) 457-8650

GLOBAL WEATHER DYNAMICS
2400 Garden Rd.
Monterey, CA 93940
(800) 538-9507
(408) 649-4500

OCEAN ROUTES/TYMSHARE
Weather Network Division
680 W. Maude Ave.
Sunnydale, CA 94086-3518
(408) 245-3600

PRODIGY
445 Hamilton Ave.
White Plains, NY 10601
(800) 284-5933

TABS/AVIOTEX
3158 Redhill Ave.
Suite 270
Costa Mesa, CA 92626
(800) 255-8227
(714) 557-9210

WSI CORPORATION
41 North Rd.
Bedford, MA 01730-0902
(617) 275-5300

Index

◆

THE MOUNTAINEERS, founded in 1906, is a non-profit outdoor activity and conservation club, whose mission is *"to explore, study, preserve and enjoy the natural beauty of the outdoors...."* Based in Seattle, Washington, the club is now the third largest such organization in the United States, with 12,000 members and four branches throughout Washington State.

The Mountaineers sponsors both classes and year-round outdoor activities in the Pacific Northwest, which include hiking, mountain climbing, ski-touring, snowshoeing, bicycling, camping, kayaking and canoeing, nature study, sailing, and adventure travel. The club's conservation division supports environmental causes through educational activities, sponsoring legislation, and presenting informational programs. All club activities are led by skilled, experienced volunteers, who are dedicated to promoting safe and responsible enjoyment and preservation of the outdoors.

The Mountaineers Books, an active, non-profit publishing program of the club, produces guidebooks, instructional texts, historical works, natural history guides, and works on environmental conservation. All books produced by The Mountaineers are aimed at fulfilling the club's mission.

If you would like to participate in these organized outdoor activities or the club's programs, consider a membership in The Mountaineers. For information and an application, write or call The Mountaineers, Club Headquarters, 300 Third Avenue West, Seattle, Washington 98119; (206) 284-6310.

ABOUT THE AUTHOR: Born and raised in the Midwest, JEFF RENNER spent his youth devouring books about mountain adventure, and took every opportunity to ski "out West" during his college days. During his first job for Seattle's KING TV, he gained an intimate knowledge of mountain weather as an on-the-scene science reporter before and during the eruption of Mount St. Helens. Since then, Renner has spent a large part of his life in and above the mountains as a skier, climber, hiker, and flight instructor. As meteorologist for KING TV, he endeavors to improve the quality and scope of weather information for all who work or play in the great outdoors.